ICT - ENERGY CONCEPTS FOR ENERGY EFFICIENCY AND SUSTAINABILITY

Edited by **Giorgos Fagas, Luca Gammaitoni, John P. Gallagher** and **Douglas J. Paul**

ICT - Energy Concepts for Energy Efficiency and Sustainability

http://dx.doi.org/10.5772/62522
Edited by Giorgos Fagas, Luca Gammaitoni, John P. Gallagher and Douglas J. Paul

Contributors

Vincent Heuveline, Manuel F. Dolz, Thomas Ludwig, A. Cristiano I. Malossi, Enrique S. Quintana Orti, M. Reza Heidari, Martin Wlotzka, Luca Gammaitoni, Kerstin Eder, Azam Seyedi, Giorgos Fagas, John P. Gallagher, Douglas J. Paul, Dirk Pesch, David Boyle

CBS, Edition **2018**

Published by InTech

Janeza Trdine 9, 51000 Rijeka, Croatia

Publishing Process Manager Ana Simcic

Technical Editor SPi Global

Cover Designer InTech Design Team

Additional hard copies can be obtained from orders@intechopen.com

ICT - Energy Concepts for Energy Efficiency and Sustainability, Edited by Giorgos Fagas, Luca Gammaitoni, John P. Gallagher and Douglas J. Paul
p. cm.
Print ISBN 978-953-51-3011-6
Online ISBN 978-953-51-3012-3

Contents

Preface

The energy consumption from the expanding use of information and communications technology (ICT) is unsustainable with present drivers, and it will impact heavily on the future climate change. However, ICT devices have the potential to contribute significantly to the reduction of CO_2 emission and enhance resource efficiency in other sectors, e.g. transportation (through intelligent transportation, advanced driver assistance systems and self-driving vehicles), heating (through smart building control) and manufacturing (through digital automation based on smart autonomous sensors). To address the energy sustainability of ICT and capture the full potential of ICT in resource efficiency, a multidisciplinary ICT-Energy Community needs to be brought together covering devices, microarchitectures, ultralarge-scale integration (ULSI), high-performance computing (HPC), energy harvesting, energy storage, system design, embedded systems, efficient electronics, static analysis and computation.

In a previous volume (*ICT-Energy-Concepts Towards Zero-Power ICT*; referenced below as Vol. 1), we addressed some of the fundamentals related to bridging the gap between the amount of energy required to operate portable/mobile ICT systems and the amount of energy available from ambient sources. The only viable solution appears to be to attack the gap from both sides, i.e. to reduce the amount of energy dissipated during computation and to improve the efficiency in energy-harvesting technologies. In this book, we build on those concepts and continue the discussion on energy efficiency and sustainability by addressing the minimisation of energy consumption at different levels across the ICT system stack, from hardware to software, as well as discussing energy consumption issues in high-performance computing (HPC), data centres and communication in sensor networks.

Bringing the community together is a journey, and it requires the dedicated contribution of a multidisciplinary team of scientists, engineers and technologists that are actively involved in forefront research in the growing ICT-Energy field. The journey starts in Chapter 1 with a general overview of the challenges and opportunities in this emerging field and a common framework to strive towards energy sustainable ICT. Chapter 2 then takes us to an indispensable introduction to the underlying principles of energy considerations introduced in later parts of this volume. Building on the fundamental concepts of energy (Chapter 2 of Vol. 1), Luca Gammaitoni takes the narrative further by introducing the fundamentals of energy transformations (and losses in the form of heat) in information processing. Whereas defining the energy as the 'capability of performing work' serves well to introduce the energy concept, it is much harder to find ways of accurately quantifying it. In Chapter 3, Kerstin Eder and colleagues take the challenging task of introducing methods to measure the energy consumed by software instructions on target hardware platforms. Starting from introducing the fundamentals of energy measurement techniques, specific examples of measurements on real applications are discussed to make the learnings concrete. Guidelines are also provided on what needs to be considered when an energy measurement system is designed. The rest of the volume slowly delves deeper on issues.

To address energy efficiency and sustainability in ICT, energy optimisations across the whole system stack are needed, and one of the key issues is to derive benefits from new devices in circuit architectures. In Chapter 4, Osman Ünsal and colleagues describe a new design paradigm for a memory cell based on novel switches and standard CMOS. The authors' design provides a specific example of hardware considerations, therefore offering an illustration of the 'ICT-Energy Concepts' that need to be considered by the community. All the concepts needed to understand the target design are introduced, therefore educating the reader in the process.

In Chapter 5, using the lenses of a programmer, Kerstin Eder and John P. Gallagher introduce energy efficiency as a software engineering design goal and discuss the concept of energy transparency to narrow the gap between the best efficiency a hardware platform is capable of and the efficiency achieved when running a particular set of applications. After a powerful motivating introduction on energy-aware software engineering, tools such as energy modelling and static energy analysis are presented in a pedagogic way in support of the energy transparency concept from the software perspective. The link to energy information about the hardware is also discussed. The generic approach adopted by the authors can be applied on a variety of hardware platforms (high-performance systems, smart sensor deeply embedded systems, mobile devices, data centres) independent of a particular class of applications or programming languages.

The authors of Chapter 6 take the task of discussing energy-aware software principles in the context of high-performance computing, one of the major energy consumers. Discussing the specifics of solving algebraic systems of equations, Martin Wlotzka and colleagues present the core issues of HPC algorithms and follow with an analysis of energy measurements that can be used to characterise and optimise their energy consumption. They demonstrate through specific examples how tools can be devised to classify and predict the energy and performance costs of sparse linear algebra operations which are at the heart of scientific (and not only) computing. Although some of the dense mathematics may at first alert the uninitiated reader, the chapter flow is such that at the end the authors' objectives to familiarise us with the issue and the challenges involved are achieved.

It is foreseen that an ever-increasing number of intelligent, mobile, sensing and communicating devices will be dispersed into ordinary appliances and tools of common use. Most wireless network applications require embedded devices to be miniaturised and energy efficient. However, one of the key issues is the energy consumption associated with wirelessly communicating the data of the embedded systems within the network. This is a long-standing problem that David Boyle, Roman Kolcun and Eric Yeatman take the task to introduce Chapter 7. While offering a detailed accurate account of the development of the field and its current status including standards, the authors also explain the necessary trade-offs in the design space for a variety of traffic patterns, particularly at the link layer and routing protocols.

In Chapter 8, Dirk Pesch and colleagues discuss the architecture and implementation of an integrated energy management system for data centres which is capable of up to 40% savings in total energy consumption. The proposed system combines a number of ICT-Energy Concepts discussed in this book and in Vol. 1. These include the use of sensor networks, thermal management measurements and modelling and actuation for the monitoring and control of IT workload, data centre cooling, local power generation and waste heat recovery. The authors also describe how the architecture of the platform is designed so that the system interacts with the simulated data centre in the same manner as it interacts with the components in a real data centre. This is a very powerful concept for the testing and commissioning of novel management and control systems before their target deployment in real life.

The volume concludes with a complementary discussion on energy generation from ambient sources. In Chapter 9, Lourdes Ferre Llin and Douglas J. Paul discuss the use of renewable energies in powering ICT devices and systems, thereby achieving energy sustainability. Particularly, the authors discuss the conversion of heat and light into electricity through, respectively, the thermoelectric and photovoltaic effect. Starting from an introduction to the basic physics of the two effects and the operation principles of both types of sustainable energy sources, the authors introduce the fundamental limits of both technologies and discuss their application in the deployment of ICT devices and systems. A number of factors are important so that the required energy and power can be provided by heat and light, and the authors achieve to strongly supplement the discussions in Vol. 1.

Finally, this book was realised thanks to the contribution of the project 'Coordinating Research Efforts of the ICT-Energy Community' funded from the European Union under the Future and Emerging Technologies (FET) area of the Seventh Framework Programme for Research and Technological Development (grant agreement n. 611004). My greatest appreciation and gratitude go to our colleagues, all leaders in their respective field, who invested their time in bringing these issues to our attention. In this journey, we have many more places to discover, and it is my hope that their contributions will provide students, young research scientists and the broader community with food for thought until the next stop.

Giorgos Fagas
Tyndall National Institute - University College Cork
Cork, Ireland

Chapter 1

Energy Challenges for ICT

Giorgos Fagas, John P. Gallagher,

Luca Gammaitoni and Douglas J. Paul

Additional information is available at the end of the chapter

http://dx.doi.org/10.5772/66678

Abstract

The energy consumption from the expanding use of information and communications technology (ICT) is unsustainable with present drivers, and it will impact heavily on the future climate change. However, ICT devices have the potential to contribute significantly to the reduction of CO_2 emission and enhance resource efficiency in other sectors, e.g., transportation (through intelligent transportation and advanced driver assistance systems and self-driving vehicles), heating (through smart building control), and manufacturing (through digital automation based on smart autonomous sensors). To address the energy sustainability of ICT and capture the full potential of ICT in resource efficiency, a multidisciplinary ICT-energy community needs to be brought together covering devices, microarchitectures, ultra large-scale integration (ULSI), high-performance computing (HPC), energy harvesting, energy storage, system design, embedded systems, efficient electronics, static analysis, and computation. In this chapter, we introduce challenges and opportunities in this emerging field and a common framework to strive towards energy-sustainable ICT.

Keywords: ICT, energy efficiency, energy sustainability, low power, embedded systems, smart sensors, high-performance computing, data centres, Internet of Things

1. Introduction

The reliance of society on the use of information and communications technology (ICT) devices and systems is ever increasing. From the proliferation of e-mail and electronic document exchange, social media and apps to the ready use of mobile devices (already in their fourth generation), data analytics, and advanced computing to solve big challenges, there has been a transformative impact on society. However, the expanding ICT use requires increasing amounts of electricity to run and it implies fundamental transformations of energy that result in energy lost in the form of heat as explained in Chapter 2. Several models exist on the energy/electricity consumption of ICT, and some of them will be referenced below and in the remainder of the book. Nevertheless, a conservative estimation currently puts around 4% of all electricity consumption and over 2% of all CO_2 emissions as the result of ICT use. If entertainment, telephones, TV, and media that are now being translated onto ICT devices and systems are added, then these consumption numbers approximately double. In a recent study, the share of ICT global electricity usage by 2030 was estimated at 21% in a likely scenario and 51% in the worst-case [1].

By any account, the increasing energy consumption and the associated CO_2 emission of ICT devices strain the targets of low carbon, resource efficiency, and competitiveness of any modern circular economy. At present, all ICT roadmaps still use cost or performance as the main driver and improved energy management as a secondary issue, i.e., energy issues such as production, efficiency, and storage are considered only if necessary to achieve cost reduction or performance enhancement. However, if ICT is to become sustainable in terms of energy, then energy must be the key driver for all ICT devices and systems. Sustainable energy was defined by the United Nation's Brundtland Commission "Our Common Future" in 1987 [2] as requiring fuel or energy sources that have the following criteria: fuel is not significantly depleted by continuous use; no significant pollution or hazards to humans, ecology, or climate systems; no significant perpetuation of social injustice.

There are two main aims that must be achieved in order to meet this vision. The first is that the consumption of energy by all ICT devices and systems must be reduced. The second is that the use of sustainable energy and, in particular, renewable energy systems must be increased to power the majority of ICT. In fact, these aims represent also strategic conditions for the future development of ICT itself:

- **High-performance computing systems** are the ICT-enabling technology for advanced mathematical modelling and numerical simulations that play a key role in scientific discovery and technological innovation. If we want to foster the realization of the next generation of high-performance computing (HPC), we need to increase energy efficiency of computing. Exascale computers capable of reaching 10^{18} operations per second require a substantial decrease in the amount of energy dissipated into heat compared to present standards. There is a significant drive for energy efficiency in computing architectures, both for designing next-generation hardware, from the fundamental devices of information processing to data storage architecture and communication networks, and for developing software tools and algorithms to increase the efficient use of the hardware.

- **Smart autonomous sensor systems** for the so-called Internet of things (IoT) scenario. IoT foresees that an ever-increasing number of intelligent, mobile, sensing, and communicating devices will be dispersed into ordinary appliances and tools of common use. But most applications require an IoT device to be miniaturized, energy-efficient, and autonomous so that it is portable and self-sustaining. To achieve this, the amount of energy required by such devices needs to be significantly reduced and conventional power management needs to be replaced with energy-saving devices and other methods to regulate power

http://dx.doi.org/10.5772/66678

supply and demand. Emerging autonomous sensors need to maintain ultra-low power (ULP) duty cycles and incorporate an energy harvesting source, an energy storage device, and electronic circuits for power management, sensing, and communication into sub-cm scale systems.

- **Data centres** have become a critical ICT infrastructure owing to software as a service, mobile cloud applications, digital media streaming, and the expected growth of IoT. Data centre energy consumption is currently growing at a compound annual rate of over 10%. Power and thermal monitoring and control as well as recovery of waste heat play a key role in reducing consumption and economic costs. However, ever more opportunities exist towards a comprehensive integrated energy management system to enhance the energy and power management of data centres in conjunction with renewable energy generation and integration with their surrounding infrastructure.

To bridge the gap between the energy and power requirements in ICT and availability from energy harvesting/renewable sources, a multidisciplinary effort is required to address energy and power management issues across the layers of ICT systems. Several concepts for energy efficiency and sustainability are discussed in the forerunner of this book (to be referenced as Vol. 1) [3] and in the other Chapters. In Section 2, we discuss the energy sustainability of ICT, and in Section 3, we present challenges and opportunities in the framework of the ICT system stack.

2. The energy and power issue for ICT

Energy is the "capability of performing work." Power, being equal to the work divided by the time that it takes to do it, indicates how fast this work is done for a certain amount of energy or how much energy is consumed to achieve a task, e.g., a computational operation (in a specified time interval). In Chapter 2 of Vol. 1, the concept of energy and its relationship to power is discussed. In the next chapter of this book, the fundamentals of energy transformations (and losses in the form of heat) in information processing are discussed. An introduction to measuring energy consumption in computing is provided in Chapter 3. However, in anticipation to the discussion below and the general theme of the book on ICT-Energy issues, it should be understood that apart from reducing consumed energy there is also a balance to be struck with the ability to perform critical tasks. For the former, energy-efficient computing architectures could minimize energy-loss operations in information processing, data transfer and communications. On the other hand, taking the example of a wearable glucose sensor does require specified amounts of energy to be available at regular intervals so that monitored data are transmitted over the necessary distance via an antenna. Bearing these considerations in mind, the power required to operate current ICT systems ranges from the mW level for small autonomous sensor systems to tens of MWs for HPC systems. In between these power levels lie a large number of devices including embedded sensors, mobile phones, smartphones, tablets, personal computers, servers, and cloud computing storage systems. The annual sales of many of these consumer systems are now at the 100 million to 1 billion per annum and, without taking into account the energy required for production, every device consumes a certain amount of energy that results in the emission of CO_2.

The ubiquitous use of ICT systems has been driven by the continuous scaling of silicon chips. The original drivers for scaling came from improving the performance of computers, but as the size of transistors was reduced, so was the power consumption enabling many portable systems to be developed. **Figure 1** presents a summary of the scaling of silicon chips and demonstrates how the performance of computers has improved over time along with the number of transistors physically produced on each chip. However, as the number of transistors increased according to constant electric field scaling (Dennard's scaling rules [5]), the device sizes started to become so small for significant quantum effects such as tunnelling to kick in. While previously integrated circuit technology (based on complementary metal-oxide-semiconductor—CMOS) was dominated by dynamic power dissipation, leakage of currents due to quantum tunnelling started to become significant for advanced scaled devices. Fred Pollack from Intel first suggested the problems of continuing the scaling of the 1990s without changes to the architecture where the chip power density would scale to that of a nuclear reactor [6]. Indeed in 2006, Intel released a chip with a power density higher than the core of a nuclear pressurized water reactor (PWR). Chip design was changed to reduce this power density (**Figure 2(a)**), but an analysis of the absolute power indicates that with the increasing number of transistors, the total power of microprocessor units (MPUs) is still increasing over time despite the reduction in power density (**Figure 2(b)**). More worrying is the increase in peak power dissipation of low power, portable MPUs that are being driven by applications such as video streaming. This has led to architectures such as the ARM Big.Little [8], which during normal operation can allow low power operation, but when computationally intensive tasks must be undertaken, the peak power will increase significantly as required by the video streaming applications for compression and decompression.

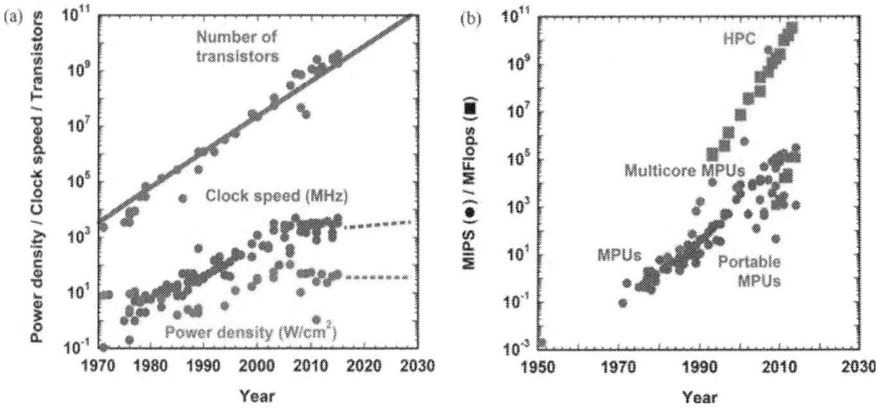

Figure 1. (a) The scaling of the number of transistors, chip clock speed, and power density as a function of time. (b) The performance of microprocessor units (MPUs) for computers, portable devices, and high-performance computing system as a function of time. The circles are data for million instructions per second (MIPS), while the squares are million floating point operations per second (MFlops) (Sources: Datasheets for processors from Intel, AMD, IBM, Digital, Motorola, Zilog, Samsung, Apple and Top 500 HPC [4]).

http://dx.doi.org/10.5772/66678

There are a large number of market surveys predicting the future of the ICT market and all of them suggest growth in a significant number of areas. **Figure 3(a)** shows the increase with time of the total number of ICT devices being sold each year. Only standard PCs and set top boxes are predicted to be static or decrease, while all other areas are predicted to grow significantly, suggesting that the number of ICT devices will continue to grow in the foreseeable future. As the number of ICT devices increases, and especially with the use of portable devices and the proliferation of the IoT, the amount of data being transmitted by the Internet and communication networks is also increasing significantly (**Figure 3(b)**).

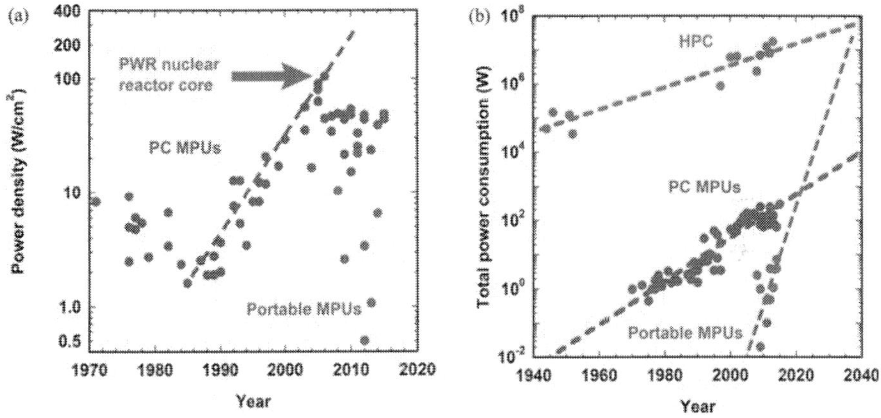

Figure 2. (a) The power density for PC and portable MPUs versus year of first release. The PWR nuclear reactor core power density is 102 W/cm² [7]. (b) The total power consumption of HPCs, MPUs for PCs (servers, desktop, and laptops), and MPUs portable devices (smartphones and tablets). Datasheets for the processors from Intel, AMD, IBM, Digital, Motorola, Zilog, Samsung, Apple, and Top 500 HPC [4].

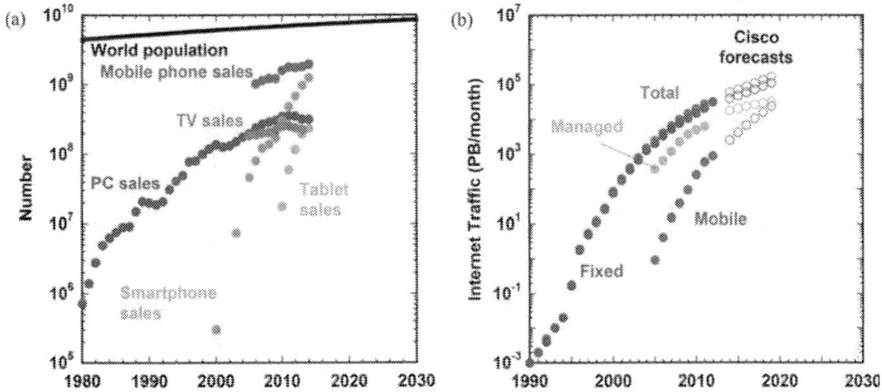

Figure 3. (a) The number of shipped end user device products per annum for personal computers (PCs—both desktops and laptops), mobile phones, smartphones, tablets, and TVs (Sources: [9, 10]). (b) The average Internet traffic in terabytes per month for each year (Source: [11]).

A number of studies have been looking at the energy consumption of ICT devices and systems (**Figure 4**). While several sources especially on web pages provide guesses of the total electricity consumption of ICT devices, there are a number of detailed studies that have tried to accurately estimate the total energy consumption and CO_2 emission [12–15]. These studies have used trade data to estimate the number of devices, analyzed average use and loading of devices and considered the power scaling to provide estimates of the total electricity consumption (**Figure 4**) and CO_2 emission (**Figure 5**). In particular, there are a significant number of studies investigating the energy consumption and scaling of the Internet as the present energy consumption of telecommunications is the fastest growing part of ICT energy consumption. The key message is that in 2015, around 4% of the electricity generated worldwide is consumed by ICT devices [16], which results in 1 billion tonnes CO_2 equivalent, that is, about 2.3% of the global emission of CO_2. These studies do not include TV and media uses of ICT systems or the CO_2 produced from the manufacture of the ICT devices and systems. The suggestion from reference [14] is that TV and media has 82% of the energy consumption of ICT but 131% of the CO_2 emission of ICT. As media is now being transferred to ICT devices and systems, a large part of this consumption may require being included in the total ICT consumption and emission in the future.

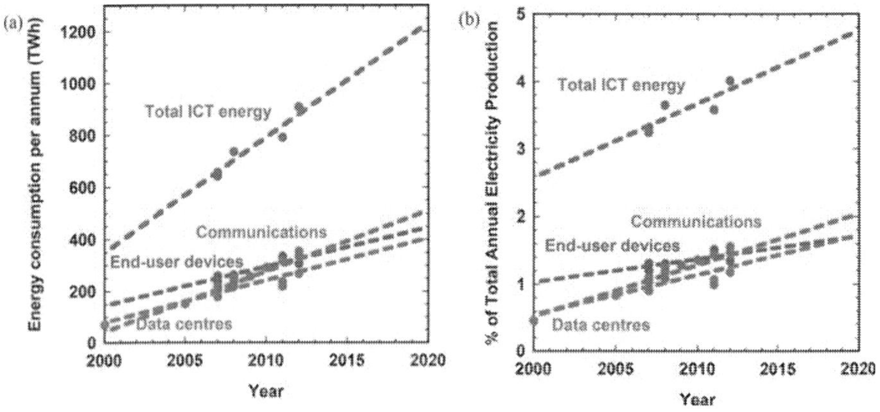

Figure 4. (a) The estimated energy consumption per annum for data centres, PC devices (including desktops, laptops, and tablets), communications (Internet, networks, mobiles, and smartphones), and data centres (servers including cloud computing) plus the total annual ICT energy consumption. Total ICT energy does not include any entertainment and media use such as TV, HiFi, DVD, CD, or radio. The ICT energy also excludes all manufacture and disposal of ICT devices (Sources: [12–15]).

An obvious way to reduce CO_2 emissions is to use sustainable and in particular renewable energy generation sources. **Figure 6** provides a comparison between the power requirements of ICT devices and systems and available sustainable energy generation technology. For a large-scale energy generation, the real issue is that most sustainable energy technologies require significantly higher capital outlay for installation (e.g., photovoltaic and hydro) and have long payback periods. Also many of the renewable sources cannot deliver a constant supply of energy and require both storage and/or alternative power supply mechanisms to maintain a

http://dx.doi.org/10.5772/66678

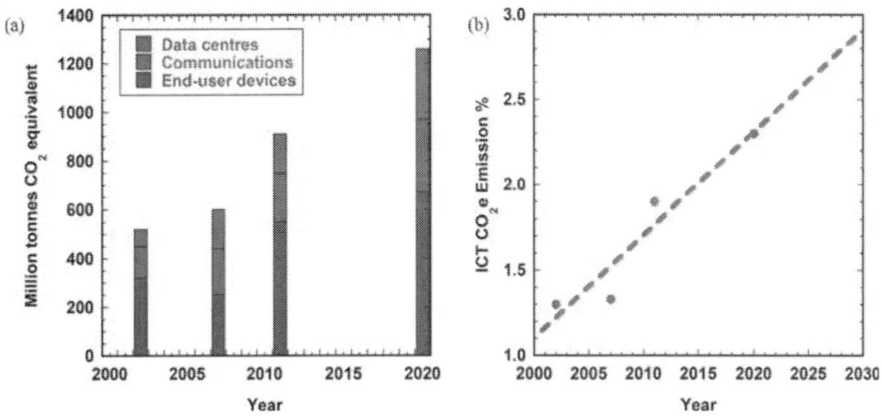

Figure 5. (a) The amount of CO_2 equivalent emitted from the manufacture and use of ICT equipment, infrastructure, and systems per annum along with predictions for 2020. (b) The percentage of ICT CO_2 equivalent emissions as a percentage of total CO_2 emissions (Sources: [14, 15]).

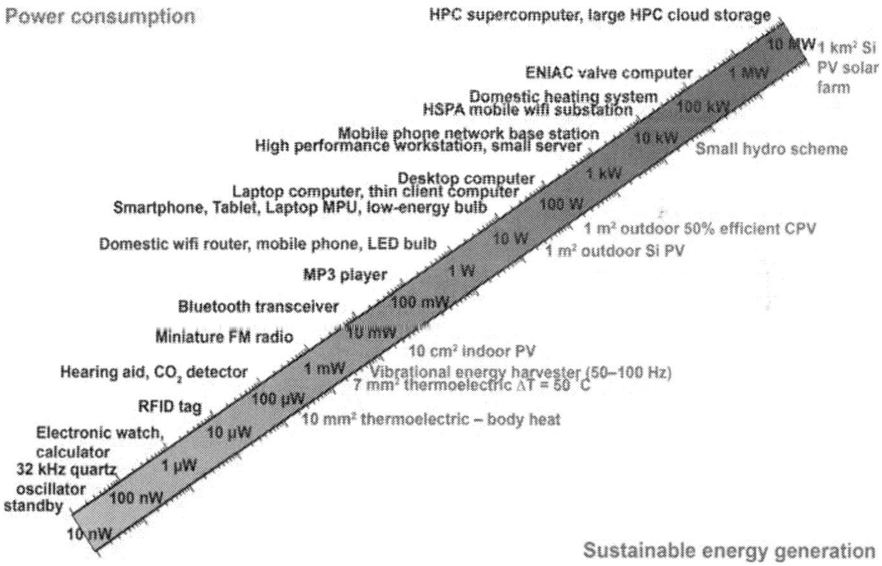

Figure 6. Typical power consumption of different ICT devices and systems versus the power that can be generated from sustainable/renewable energy generation devices and systems.

constant energy supply. At the small scale, batteries and super capacitors could potentially deal with such issues with advances in power management and disruptive technologies in micro-energy storage and harvesting. At the large scale, for HPC and cloud computing, the storage of large amounts of energy is problematic and only pump-storage hydro can reach the required volume of energy in a sustainable manner. Such hydro schemes can only be built in suitable environments where large reservoirs with significant height difference can be built, and the location of such environments is seldom where the energy is required. Compressed air energy storage (CAES) is another suitable technology for bulk energy storage, but it requires underground caverns so it is also site-specific. An emerging technology that can address those issues is liquid air energy storage (LAES). LAES is modular and site agnostic, so it can be utilized for any size of energy storage and power rating up to several MWs. For specific applications, the advances and cost reductions in certain types of batteries (especially for flow redox and Li-Ion) also qualifies them as competitive technologies for significant energy storage.

3. Sustainable energy ICT: science/technology issues and opportunities

The system stack shown in **Figure 7** is a useful pictorial representation of the whole ICT system. While one might expect data centres (cloud) and HPC and smart autonomous sensors to be composed of completely different subsystems/layers, the system stacks are very similar in many areas and provide an opportunity to learn how to improve each from the experience of the other.

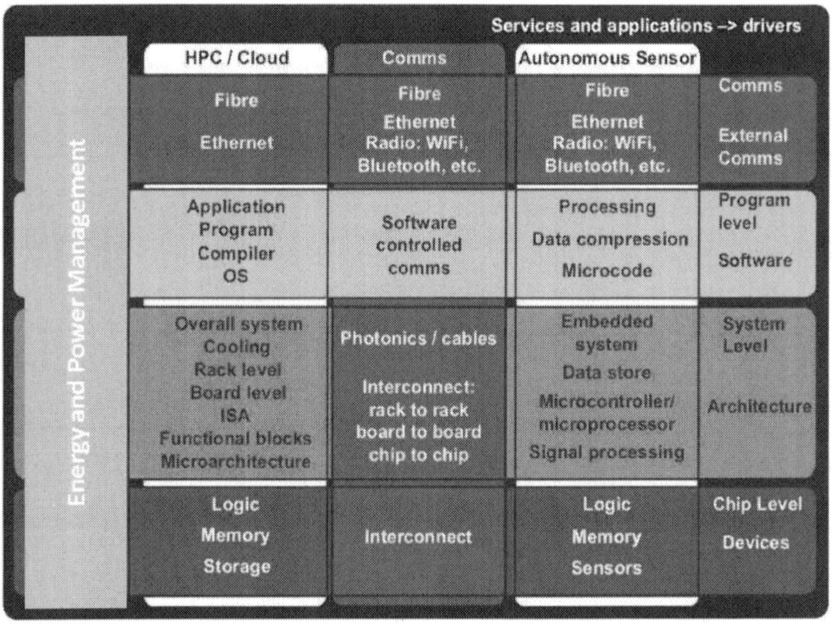

Figure 7. A schematic diagram of the system stacks for high-performance computing (HPC), cloud computing (Data Centre), and smart autonomous sensors. The figure indicates the common elements in the system stacks between these different ICT systems.

http://dx.doi.org/10.5772/66678

3.1. Chip level (devices)

From switches for logic and memory cells to transducing elements, devices provide the fundamental information processing. A generic ICT device can be viewed as a machine that processes information while transforming work into heat and heat into work. Pioneering research developed by J. Von Neumann and by R. Landauer in the last century has shown that information processing is intimately related to energy management [17]. An ICT device (see **Figure 8**) is a machine that inputs information and energy (under the form of work), processes both and outputs information and energy (under the form of heat). From this perspective energy dissipation via heat production and energy transformation processes are two aspects of the same topic: energy management at the micro- and nanoscales. For our purposes, energy efficiency is defined as the percentage of energy input to a device consumed in useful work and not wasted as heat. This definition, however, may not apply when we have to deal with processes taking place at the nanoscale (see Chapter 2 of Vol. 1).

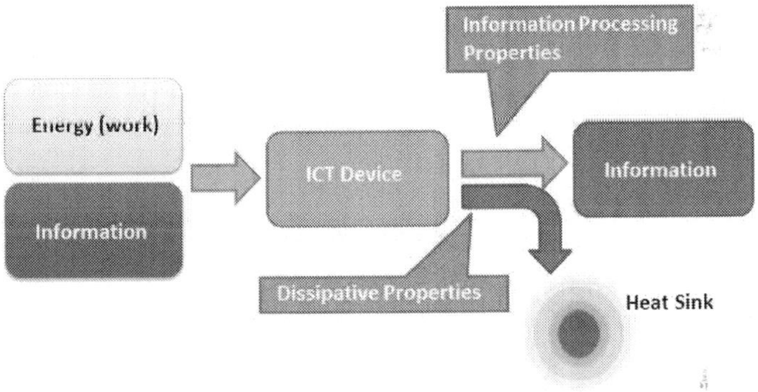

Figure 8. An ICT device is a machine that inputs information and energy (under the form of work) and processes both and outputs information and energy (mostly under the form of heat).

In sensor systems where conversion of external stimuli to signals is required, there exist several options of low-power transducer devices based on micro-/nanoelectromechanical systems (MEMS/NEMS) as well as optical and electrochemical sensing mechanisms. However, energy considerations become significant at the next stage that requires analyzing inputs and performing computational operations. As mentioned earlier, the side effect of the advances in the computation process is the increasing heat production. Here the workhorse has been the field effect transistor (FET), and in the last 40 years, the semiconductor industry has made impressive progresses in reducing the size of the CMOS components, thus increasing the computational density of microprocessors. New types of scaling rules as well as new designs and materials were introduced to reduce the energy dissipation following the breakdown of the Dennard scaling rules (where a number of fixed parameters in the transistor did not scale in a standard linear, quadratic or cubic way with the gate length of the transistor, e.g., the voltage threshold at which the transistor is switched on into a digital "1" state).

While in principle, the switching energy of a metal–oxide–semiconductor field-effect transistor (MOSFET) could be reduced by reducing the supply voltage V_{DD} (indeed, this is what scaling over the last 40 years has been doing to reduce the energy and power dissipation), the finite bandgap of a semiconductor provides a lower limit below that the transistor will not switch on. Moving to another semiconductor with a smaller bandgap (than silicon) allows some improvement to reduce energy, but (apart from technological constraints) the fundamental physics of the p-n junction for the contacts provides another limit for switching MOSFETs on and off. A simplified analysis of a generic electronic switch at thermal equilibrium, modelled as a potential barrier separating two quantum wells (an idealized FET channel between source and drain contacts) showed that the channel could conceivably be scaled down to ~1.5 nm and the transistor could have a minimum switching speed of ~40 fs [18] (for the fundamentals of minimum energy of computing see also Chapter 7 of Vol. 1 and Chapter 2 of this volume). This is significantly smaller and faster than the FETs of today. FETs, however, have fundamental power and speed limits, and these cannot be overcome even by switching to new materials. For example, in order to avoid leakage at room temperature due to a finite height of the potential barrier, the voltage cannot be scaled as rapidly as the physical dimensions and the resulting power density for these switches at maximum packing density would be on the order of 1 MW/cm^2. The steep increase in CPU power density is mirrored by the increasing fraction of energy spent in cooling activities. Whereas there is still significant progress to be achieved by advanced CMOS, only by moving to a radically new technology can lower power dissipation potentially be achieved.

Figure 9 provides a comparison of present and future high-performance and low-power CMOS against future proposed device technologies. Present low power CMOS is already at 3 aJ per operation (excluding large fan-out and interconnect impedance) and only predicted to reduce by a factor of 3 over the next 13 years. The majority of improved future technologies do not produce any significant reduction in energy consumption per operation. Indeed a 100 times the Landauer thermodynamic limit appears to be difficult to better when all the new proposed device technologies are considered. This is 100 times the Landauer limit at 300 K and is related to the energy barriers in all the switching devices that provide a significant I_{on}/I_{off} ratio as required for the circuit architectures. A number of new types of devices suggest that this barrier can be circumvented. A switch with 8 × 10^{-20} W of switching power has been demonstrated albeit at the slow switching speed of ms [21]. More importantly, this switch demonstrates that the 300x Landauer limit can be broken.

Key device challenges can be summarized as:

- The cost of the lowest energy devices requires the latest CMOS technology node that now requires enormous economies of scale due to the cost of the foundries and technology.
- The scaling of transistors to smaller dimensions is now not expected to decrease the switching energy significantly.

http://dx.doi.org/10.5772/66678

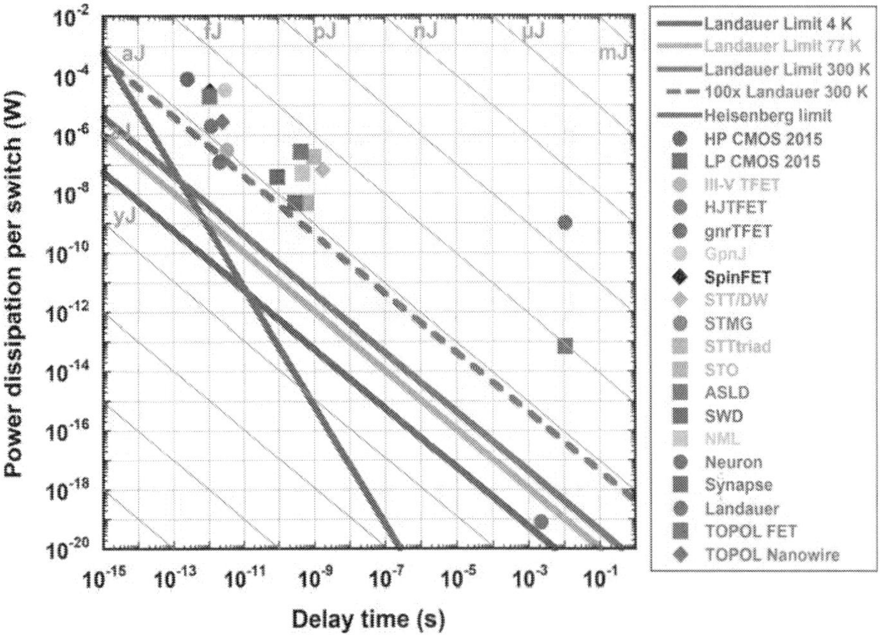

Figure 9. The intrinsic power dissipation per device switch versus delay time for CMOS and future devices as proposed in the ITRS future emerging technology roadmap [19, 20]. The diagonal grey lines are lines of constant energy. The figure indicates a limit of about 1 aJ per switch for CMOS devices and 250 zJ per device for the best proposed future switching device. Key: CMOS HP (high-performance CMOS), CMOS LP (low-power CMOS), iii–vTFET (III–V tunnel FET), HJFET (heterojunction FET), gnrFET (graphene nanoribbon TFET), GpnJ (graphene p-n junction), spinFET (Sughara-Tanaka spin FET), SST/DW (spin torque domain wall), STMG (spin torque majority device), STTtriad (spin torque triad), STO (spin torque oscillator), ASLD (all spin logic device), and NML (nanomagnetic logic). Results from the MINECC projects are the Landauer MEMS device (Landauer) [21], the Si CMOS FET from TOPOL (TOPOL FET) and the Si nanowire from TOPOL (TOPOL Nanowire).

- If there is a change of the basic switching device to move to a significantly lower energy technology, then circuit architectures, design tools, verification, operating systems, and software may require rewriting or complete changes for basic operation or optimal performance.
- Driving interconnects with multiple fan-out or an antenna have fundamental energy and noise limits that makes ultra-low energy consumption difficult.

Significant opportunities can be divided into continued scaling of conventional transistor devices (termed More Moore) and adding new device concepts for computation (termed beyond Moore or beyond CMOS) or new functionality onto the base CMOS technology (termed More than Moore). The key concept in More Moore is that the circuit architectures will be similar to present CMOS architectures and only the devices, voltages, or currents are changed. More Moore is looking at the challenges of scaling CMOS devices to dimensions below 10 nm, and a significant portion of this work is now investigating new channel materials and device architectures such as gate-all-around nanowires that will allow higher performance with lower voltages for reduced power operation. The materials include Ge, III–Vs, semimetals, carbon-based materials, magnetic materials, and phase-change materials. Steep sub-threshold slope transistors allowing voltages for operation below the normal p-n junction limits are also being investigated. The major issue now being recognized by the semiconductor industry association and others is that scaling the transistor to smaller sizes may no longer result in lower energy devices; hence, More than Moore increased functionality is now a very diverse field of study:

- Devices with learning capability, e.g., devices that learn logic configurations and devices that learn by example.
- Integration of RF devices and components.
- Optical devices, e.g., Si photonics for communications and optical sensing modalities.
- Micro electro mechanical systems (MEMS) and nano electro mechanical systems (NEMS).
- Integration of sensors.

Beyond Moore is the opportunity to completely change how information is processed. This field is investigating the fundamental limits of information theory and if any of the known limits can be circumvented through innovative techniques. One example includes investigating if the Landauer limit requiring heat to be dissipated through the storage and erasure of information can be circumvented to allow zero energy switching. Spin wave devices are another example for a radical new technology that circumvents many of the limits of conventional transistor logic. Investigating methods of integrating device operation with energy harvesting has also been suggested where the heat dissipated and normally lost could be harvested to improve the overall thermodynamic system efficiency. The Beyond Moore research area could have a large impact in reducing energy consumption of ICT devices but is also the most difficult to implement into systems as solutions may be radically different from conventional CMOS technology, architecture, and systems.

3.2. System level (architecture)

3.2.1. Circuit microarchitecture

The microarchitecture level considers the integration of many of the underlying fundamental device technologies. Transistor technologies and silicon processes are combined into useful blocks such as memory (see, e.g., Chapter 4), control, and arithmetic that together form a computational device, often a processor or an accelerator. The scope for that integrated device is very broad, including ultralow power embedded devices, through general-purpose processors, up to high-performance network-on-chip components. As indicated in Section 3.1, current manufacturing of semiconductor devices has hit a fundamental efficiency limit called the "energy wall" that prevents reduction of energy consumption when transistor size scales down for forthcoming technology nodes. Both at a small-scale (embedded systems) and at a large-scale (HPC/Data centres), the so-called economic meltdown trend of Moore's law [22] transcends in a dramatic increase in the computation and cooling energy costs. Based on current projections, a tenfold improvement in chip energy-efficiency is needed to maintain information technology (IT) energy scalability in the next decade. The ultimate limits from architecture designs are almost impossible to derive, but based on current technology, there is general agreement by academia and industry that new architectures are more promising to significantly reduce power consumption than improving the energy consumption of the basic switching device in the circuit.

The amount of energy consumption from a circuit architecture design for a given CMOS technology node is heavily dependent on how specific (i.e., optimized for a single or a few tasks) or how general (i.e., undertake many different computations) a design has to deliver. Applications specific integrated circuits (ASICs) designed for a single task can be optimized proving the lowest energy consumption, but such designs have no flexibility and cannot be reprogrammed. For microprocessors or microcontrollers that must be able to undertake a wide range of tasks, optimization to reduce energy consumption is significantly more difficult. Microarchitecture exploits what is physically possible with contemporary technology and presents an interface through which other hardware and software can use the device. In hardware terms, this interface is, of course, physical and will typically obey a specified protocol. In software terms, the processing device presents a set of possible operations through an instruction set architecture (ISA). If the ISA is a description of the behaviours of the device, then the microarchitecture is the implementation of those behaviours. Advances in physics, transistor design, and device manufacturing techniques can benefit microelectronic devices of all kinds; however, microarchitecture design decisions are heavily influenced by the target market of the resultant product. While all devices strive to achieve good efficiency, balancing performance and power consumption, the application area will dictate design constraints such as size, the energy budget, and maximum power. We may group microarchitecture characteristics grouped into four areas: deeply embedded, embedded/mobile, general purpose, and servers/high-performance. These are not necessarily strict boundaries and properties often transfer between areas over time, as technology or commercial pressures permit. At present, microarchitectures have significantly different drivers for HPC/data centres, general purpose (e.g., PCs), and embedded systems/portable systems.

Deeply embedded—An example of a deeply embedded device is the processor in a smart card. It must fit within a credit card form factor, cannot be modified once sent to the customer, be powered by a battery-backed device, and obey strict security protocols. A new generation of deeply embedded devices is smart autonomous sensors, which due to their ability to seamlessly integrate with the environment have given rise to cyber-physical systems (CPS) and IoT platforms. This requires integration of heterogeneous components dedicated to signal acquisition (e.g., analogue-to-digital converters (ADCs)), processing (digital signal processors), and environment manipulation (actuators). The embedded system may also typically include wireless communication and run on batteries together with energy harvesters. Hence, one or more of the following constraints may apply:

- Physical size must be small, from millimetres to a few centimetres, depending on application.
- The energy budget is finite as power may be intermittent or limited, often sub-watt or sub-milliwatt.

- Operating temperatures may vary significantly and the removal of excess heat quickly may not be possible.
- Reliability is essential, because servicing may be difficult, expensive, or impossible.
- Predictable behaviour may be required to guarantee safety criteria or always-correct device functionality.

All of the above points are relevant in an energy context. The energy consumption of the device dictates the temperature it runs at, the type of cooling required, how much processing and communication can be performed, and how long it will live. Predictable energy consumption is required to guarantee a particular battery life. To meet design goals within a small energy budget envelope, the design of each of the components is highly tailored to the targeted application (see e.g., [23, 24] and Chapter 8 of Vol. 1 for design consideration of a wireless sensor node). Circuit designs include low-voltage, power-efficient ADCs and filters, instrumentation amplifiers, domain-specific memory organizations, and schemes for energy/ power management and transfer incl. energy-efficient passives.

The microarchitectures of deeply embedded systems take various forms, but the following traits are common:

- They provide predictable execution times for many or all of the ISA instructions they support.
- Their functional blocks, such as arithmetic and memory units, are often simpler than larger counterparts, to reduce power, improve predictability, and keep the device small.
- Their memory hierarchy is flat, avoiding caches that would impact predictability and increase device complexity.
- Programs may execute directly out of integrated flash storage, with RAM only used for read-write data.

Prolific architectures in this area include AVR, ARM's Cortex M-series, and PIC. They feature compact instruction sets (often 8- or 16-bit instructions), with short execution pipelines and in-order execution. This means opportunities for performance enhancement are limited, but their implementation is simple. The relative simplicity of such devices aids activities such as modelling their behaviour in order to predict energy consumption. However, properties such as the memory layout limit the types of application that can feasibly be run on such devices.

Looking forward, we can expect the safety and real-time requirements of deeply embedded systems to persist. However, research must strive to shrink these devices into sub-millimeter and beyond, with a desire for micro- or nano-watt power envelope devices with comparable performance today, while milli-watt devices deliver improved performance and capabilities. Multi-core is not yet common in deeply embedded systems, but in order to deliver the above improvements in the light of limits to Moore scaling, we can expect it to become essential. A particularly lucrative opportunity in deeply embedded research is a "zero power" idle or

http://dx.doi.org/10.5772/66678

sleep mode, with close to instant response time. When not responding to an event, the device should, ideally, consume no energy at all. However, following an event (such as inbound sensor data), it must be able to return to a responsive state.

Embedded/mobile—The line between embedded and deeply embedded is often blurry, but for the present purpose, we group *regular* embedded devices with mobile devices, in order to set the grouping in terms of energy requirements. Such devices may still have real-time constraints, small size requirements, and sub-watt power envelopes. An example of an embedded device is the controller chip on a hard disk or a solid-state disk. They cannot be replaced, so their failure effectively renders the entire disk useless. The software running on them is difficult or undesirable to update in the interests of data security. They interact with components that have strict timing protocols, and missed deadlines will potentially result in data loss or corruption. They must fit within the form-factor of the device, e.g., the whole device may be as small as 12 × 16 mm (the smallest of the possible M.2 format of expansion devices used in laptops at the time of writing). However, the performance requirement and available power are higher than deeply embedded devices, hence, this may be considered embedded rather than deeply embedded.

Mobile and embedded systems typically have the following microarchitecture properties:

- Heavily integrated into system-on-chip (SoC), providing various peripherals and multi-core computational capabilities in a single chip.
- Larger storage and memory capacity than deeply embedded, in the order of gigabytes in current devices.
- Cache hierarchies for better memory performance.
- Sub-watt power constraints.
- Must fit within relatively small form factors, such as a mobile phone.
- More complex application sets and usage patterns.

Mobile and modern embedded devices feature performance, i.e., competitive with general-purpose hardware from less than a decade previous. A contemporary smartphone has more compute power than a ten-year-old desktop PC and consumes significantly less energy. Much of this is thanks to lower-level improvements, as described by trends such as Moore's Law and Dennard Scaling. If one is to expect the same to be true in another decade, then we must counter abstraction inefficiencies as systems become more complex. For example, May's Law states that software efficiency reduces to counteract any improvement in hardware efficiency. At the same time, as devices become even more integrated into lifestyles of the consumers, both users and app developers must be given better visibility of how and why energy is consumed by their apps. This transparency will encourage accountability for embedded software energy efficiency and narrow the gap between the best efficiency a hardware platform is capable of, and the efficiency achieved when running a particular set of applications. Energy-aware software engineering is discussed in Chapter 5.

General purpose—Global use of computing devices is shifting to a more mobile-centric approach. As such, the features of mobile devices often compete with those in more traditional general-purpose devices such as the desktop PC. However, general-purpose computing is less constrained than mobile and embedded computing, and this is reflected in the microarchitecture:

- Power envelope of tens of watts.
- Multiple layers in the cache hierarchy (three-layer caches or more).
- Less tightly integrated, with separate physical components for RAM, peripheral devices, etc.

The processor architecture may be similar to mobile devices. For example, both x86 and ARM instruction sets are used in desktop and mobile computing products. However, general-purpose microarchitectures that use these instruction sets may be more complex, with increased pipeline length, more functional units, multithreading capabilities, and higher operating frequencies. The trends in device usage suggest that general-purpose computing will merge with mobile computing, backed by servers such as those providing cloud services. To that end, the general-purpose computing category may eventually disappear, rather embedded/mobile will become the defacto general-purpose computing platform. Thus, research into low-energy microarchitecture may create more benefits if it focuses on embedded and server-related areas.

Servers/high performance—The properties of high performance and server devices are similar to general purpose, but their form factor and tolerances are different.

- Multiple servers may occupy dense racks.
- Power envelopes may be higher than general purpose due to more aggressive cooling.
- Underlying architectures are similar to general purpose, typically at leading-edge, with additional resilience features such as error correction.

Power dissipation per server must be considered in order to adequately provision the electricity supply as well as to extract the waste heat. HPC computer architectures such as server processors and MPUs, work within power constraints in the order of tens or occasionally hundreds of watts. One of the most promising lines is the development of energy-efficient processing architectures building blocks that would significantly enhance the energy-proportionality of server processing power at the deep submicron era (i.e., beyond 28 nm technology nodes). The development of these building blocks will be achieved by using emerging technologies such as fully depleted silicon on insulator (FDSOI) [25] and gate-all-around nanowires [26], and integrated microfluidic cooling and power delivery [27]. This development would require modelling the power and thermal dissipations involved in the processing units, memory hierarchy, and the cooling and power delivery networks at the server level.

Mobile architectures such as ARM are beginning to encroach into the server space, with examples such as Cavium's ThunderX processors and APM's HeliX devices, both of which use the 64-bit variant of ARM for server-grate compute and infrastructure roles. The continued proliferation of multi-core supports this movement, and it is likely we will see lower power

http://dx.doi.org/10.5772/66678

per device, but packed at a higher density so that overall the power per rack is still a challenge [28]. Interconnect between these systems also becomes of particular research interest, as the cost of data movement between these devices does not scale with the reduction in energy per processing unit.

A key challenge at the microarchitecture level is that energy efficiency advancements must take the form of a synthesis of improvements at lower levels and respond to the evolving needs at higher levels. We must strive to provide predictable, transparent behaviour, not just functionally, but for properties such as energy consumption. As discussed later, the cost of data movement must also be more readily exposed and significant effort must be put into efficiently moving (or avoiding moving) data around a system as its contribution to energy consumption will continue to increase. This must be complemented by novel, intuitive methods for presenting the increasingly heterogeneous resources that form emerging multi-core systems across the various device classes so that higher levels of the system stack can fully understand and exploit the underlying hardware. Opportunities for improvement at the microarchitecture level are along three main directions:

- Increased heterogeneity, with specialized functional blocks for particular tasks.
- Many-core systems, with various heterogeneous blocks as described above, combined with groups of homogeneous clusters.
- Network-based interconnects or complex, multilayered buses, to connect these many components, along with caches, memory, and peripherals, together.

All these pose scalability and programmability problems, some of which must be addressed at the microarchitecture level, while others can be dealt with in other levels in the stack.

3.2.2. *System architecture*

Similarly to the scenario of circuit architecture, system architecture will require multidisciplinary research efforts to achieve holistic energy-efficient design and management. Advances in computer architecture must be carried out in two information propagation directions: The first direction is bottom-up where technology-related parameters (e.g., FDSOI and Stacked-DRAMs) are propagated from new technologies and chip level to the large-scale data centre level. In particular, circuit-level technological parameters will be used in the server and data centre architecture explorations, while the chip-level cooling parameters will be exploited by the large-scale cooling and energy reuse technologies. The second direction is top-down where the target large-scale data centre operation characteristics (in terms of runtime workload variations at software level and infrastructure operating conditions), will be exploited to tune further the lower-level architectural, cooling, and technological design aspects.

In both mobile and HPC microprocessors, there are conflicting demands for high performance and energy efficiency. Circuit architectures to deliver these conflicting performance requirements are being addressed through the heterogeneous integration of a range of embedded cores. One example is the ARM big.LITTLE architecture [8], where smaller cores are employed to process simple, less demanding tasks to save energy, while larger cores are optimized for the high-performance tasks which are more energy demanding. This is a general trend, and, in parallel to it, there are a wide range of complementary techniques that are either being used in commercial processors or are being researched in academia to reduce the power consumption through architecture design. These include the following:

- *Dynamic voltage and frequency scaling*—High voltages and frequencies are used for high-performance task delivery, while low voltages and frequencies deliver reduced energy consumption [29].
- *Clock gating and clock distribution*—This is a dynamic power management technique where additional logic switches are placed between the clock and the clock input of the processors logic to disconnect the logic preventing clock cycling when it is not required [30].
- *Power domains*—These are sections of a core in a processor that can be completely powered down to reduce energy consumption without removing the supply to the system [31].
- *Pipeline balancing*—This is the dynamic adjustment of the resources of the pipeline of a processor such that it retains performance while reducing the power consumption [32].
- *Caches and interconnects*—It has been demonstrated that too large caches waste energy while too small caches limit bandwidth and performance. Careful optimization is, therefore, required to both minimize energy consumption and maintain performance [33].
- *Dynamic partial reconfiguration*—A number of field programmable gate array (FPGA) manufacturers now offer dynamic partial reconfiguration that enables reconfiguration of parts of the FPGA while the other regions are still active [34].
- *Composable and partitionable architectures*—This is where a set of low-power small cores can be aggregated together dynamically to form a larger single-threaded processor when required for higher performance [35].
- *Coarse grained reconfigurable array architecture*—Designed to be reconfigurable at the module or block level rather than at the gate level in order to trade off flexibility for reduced reconfiguration time [36].

On a larger scale, some of the new architectural solutions that are being investigated in the context of data centres and HPC include the following:

Advanced cooling infrastructures and energy reuse

- A passive thermosiphon (gravity driven) cooling system for servers and racks, using energy only to remove the waste heat out of the data centre room.
- Innovative systems able to re-convert generated heat into electricity, such as the pressure reverse osmosis (PRO) process. The objective is to provide data centres a solution to absorb a part of their waste heat and reproduce electricity.

Data centre thermal control and management

- Resolution of complex runtime multi-objective optimization (performance, power, and temperature) in high-level IT workload management (allocation and scheduling) and physical resource management (processors, memories, storage, network, and cooling infrastructure configuration).

- Exploration of several optimization schemes and control approaches such as adaptive dynamic programming, networked control, and predictive control at the large scale of data centre level.

System-level energy-efficient data centre architecture design

- Definition of more efficient system-level exploration approaches to develop energy-efficient thermal-aware architectures for servers, racks, and data centre rooms by using a holistic and tighter integration of computing and cooling power costs.
- Development of scalable simulation methods of large-scale computing systems that integrate power and thermal modelling to help data centre designers make robust predictions to anticipate potential failures of components and to accurately estimate the necessary provision in global energy during data centre conception and throughout its lifetime.

In a multiprocessor system-on-chip design (MPSoC), joint architecture optimization and integration of low-power components in novel architectures are the only way to keep increasing the performance while staying under the power budget. Solutions include application-specific and heterogeneous memory hierarchy, for instance, heterogeneous 3D architectures (stacked-DRAM) or efficient hardware implementation of components (analogue and digital) for ultra-low-power sensing. On top of this, all the on-chip components (both logic and memories) are increasingly affected by process variability, which means that they can no longer operate always under the best conditions, requiring self-adaptive architectures. Therefore, this scenario makes it more important than ever to develop scalable functional simulation frameworks to explore the limits of parallelization and global power reductions in MPSoCs. In the field of HPC and cloud computing, as volume server costs drop, electricity costs will emerge as a substantial fraction of the overall cost of ownership in HPC clusters and data centres. The fundamental question beyond the state-of-the-art is how to bridge the efficiency gap between emerging HPC, data-centric and scale-out workloads and modern server and network platforms. There is a large efficiency gap between existing server architectures and what the emerging data-centric scale-out workloads need.

Chapter 8 discuss architectural solution opportunities and implementation challenges in more detail.

3.3. Program level (software)

Software affects the amount of energy consumption in a system in a variety of ways. Possibly the most significant is that many facilities in a system operate at the request of software, meaning that processor cores and system-level consumers such as data communication or displays can consume more or less energy, depending on software algorithms. Accepted wisdom states that because of this, the best energy efficiency is achieved by having all software operate as quickly as possible, meaning that facilities can be disabled (minimizing their consumption) for longer periods. This is known as reducing static power. In addition, software also causes energy consumption through dynamic power, the cost of circuits charging and discharging as they signal digital information. Examples include the activation of different banks of memory according to access patterns, driving of data buses, and switching activity as gates in arithmetic units converge on an output. While static power tends to be controlled by the algorithm behind a piece of software, dynamic power is governed by its particular implementation. Its relation to higher-level programming constructs tends to be poorly understood.

In need of disruptive solutions, we cannot rely on a hardware delivering better energy performance in the future and so the developer must contribute to energy reduction too. An obvious barrier at present is that very few software developers have much idea of how much power their programs dissipate or which parts of a program are energy hotspots. This might even be different for the same program from one platform to another. There are good software engineering reasons for the programmer's ignorance of energy, namely, to allow programs to be ported to different platforms and to allow program design as higher levels of abstraction. The large conceptual gap from hardware, where energy is consumed, to high-level languages and programming abstractions has been created by decades of computer science research and compiler advances. Somehow, the developer using a high-level language has to understand the energy induced by the software at the hardware level, without having to measure it on a machine. A clear theme emerges that the developers of the future will rely less on the performance of processors, but instead be energy-aware. This means that they tune their software to work optimally on the available hardware, or, if feasible, choose hardware specific to the software application.

Given a program to be implemented in some programming language and a hardware platform on which it is to be executed, we may ask whether it is energy-optimal. The energy limit for a software is a relative and pragmatic one; what is the least energy a given program should consume on a given hardware platform? We assume that in answering this question, the program could be redesigned to use a more energy-efficient algorithm (for that hardware platform). Software designers and developers typically first target functionality (getting the program to do what it is supposed to do), then performance (doing it as fast as possible), and third, minimization of code development costs (productivity). Optimization of performance is highly important in some fields, especially in HPC. This usually means optimizing for minimizing the time of a computation (i.e., optimizing performance and high speed). However, there has not been significant work on optimizing the energy consumption or understanding the ultimate energy consumption limits. Energy consumption of software is not subject to fundamental limits in the same sense as hardware. A software is written using programming languages and translated into codes that are executed by hardware. This provides significant opportunities for optimizing energy consumption. Large-scale integration (LSI) logic suggests that dedicated low-power hardware circuitry may save 20%, while changes to software to better control the power states could provide power savings of a factor of 3–5. In short, more energy is wasted by software than by hardware.

In a general-purpose piece of software, reducing energy consumption will mean reducing the amount of static and dynamic power the application draws. This process requires two distinct steps: that of identifying the sources of each kind of power, and then reducing that amount of consumption. Identifying static power consumption corresponds to evaluating the amount of time taken for the program to execute, while dynamic power requires either analysis of the machine instructions that the program compiles to, or some (more or less approximate) model

http://dx.doi.org/10.5772/66678

of the energy usage of higher-level software. Both of these techniques are complex, and not immediately available to modern software developers, limiting their ability to make decisions to reduce energy. The key research challenge is to bridge the conceptual gap and make energy consumption transparent through the layers. When programmers are energy-aware, there are a number of measures that can be taken to optimize for energy efficiency of software [37]:

- Choose the best algorithm for the problem at hand and make sure it fits well with the computational hardware. Failure to do this can lead to costs far exceeding the benefit of more localized power optimizations.
- Minimize memory size and expensive memory accesses through algorithm transformations, efficient mapping of data into memory, and optimal use of memory bandwidth, registers, and cache.
- Optimize the performance of the application, making maximum use of the available parallelism.
- Take advantage of hardware support power management.
- Select instructions, sequence them, and order operations in a way that minimizes switching in the CPU and datapath.

The translation and execution process is separate from the software and depends on tools such as compilers, interpreters, and schedulers and the hardware platform. All of these may influence its energy consumption and can be varied independently of the software itself.

Below are some of the opportunities to be explored regarding software execution:

- *Energy-aware algorithms*—Already today and increasingly important in the future, algorithms need to be able to respect power and energy constraints. Power and energy need to be added to the conventional design goals of performance and correctness, and metrics for assessment need to be established. Standard interfaces and application programming interfaces (APIs) for collecting power and energy information have to be developed, supported by accurate measurements through built-in hardware counters and sensors or external measurement devices.
- *Avoiding data transfer*—Data transfer at all levels including local memory to cache, within local memory, read/write to storage, and transfer through the network is expensive both in terms of time and energy compared to floating point operations. Algorithms need to be investigated that reduce the need for data transfer and which exploit and improve data locality.
- *Data compression*—The volume of transferred data can be reduced if the data is compressed. However, the compression itself is an additional computation that consumes time and energy. Algorithms need to be investigated for their feasibility to use data compression, and the trade-off with time and energy needs to be taken into account.
- *Multiple precision algorithms*—Not all applications require the full IEEE-754 double precision accuracy, which is however often used by default. Algorithms need to be investigated for their feasibility to use multiple, lower precision data formats. This might speed up the computation, reduce the memory requirements, and reduce the data transfer, which all can

contribute to a reduced time and energy consumption. However, the numerical properties must be carefully considered since multiple precision algorithms might experience different numerical stability.

- *Reducing synchronization*—In most algorithms, at some point, the computation must be synchronized across the machine that usually imposes waiting and idle times. One example of a global synchronization that appears in myriads of algorithms is the computation of the dot product, where all processors need to provide a local contribution and the aggregate result is distributed. Research is needed for restructuring of existing and development of new algorithms that reduce the synchronizations.
- *Randomization and sampling algorithms*—Another approach to reduce synchronization and data transfer is the use of randomized or sampling-based algorithms. Such algorithms work on decoupled, independent subparts or instances of the problem and synchronize only locally or occasionally. However, such algorithms may show nondeterministic behaviour, or even sometimes fail to return a result. Research should address both the modification of existing deterministic algorithms towards randomization and sampling, and the developments of new algorithms. The numerical properties and the suitability for specific applications need to be considered.
- *Adaption to load imbalance*—Ill-balanced codes can infer substantial penalties on performance and energy consumption. Load imbalance may occur even for initially well-balanced simulations due to different numerical properties of the problem evolving in different temporal or spatial domains, after recovery from failure, or imposed by energy management.
- *Scheduling and memory management*—In order to efficiently use the massively parallel and heterogeneous platforms, power and energy-aware runtimes, scheduling, and memory techniques are required.
- *Autotuning algorithms*—Complementing the scheduling and memory management on the runtime level, algorithms need to be able to detect and tune themselves to the architecture. Numerical software libraries need to become able to choose the most efficient variant of an algorithm for a particular hardware and a particular application in an automatic way.

The lower energy bound limitation stems from the generality of hardware. The main benefit of software is that it is reconfigurable: one can take a general-purpose processor and deploy many different applications to it, perform field updates, and otherwise alter behaviour without altering hardware. This necessitates that the processor is substantially more generalized than the application, to allow for reconfiguration. It must have a large enough set of behaviours (i.e., be Turing complete) to be programmed, as well as having sufficient performance and resources to meet application budgets for time and space. This leads to a significant capability gap between hardware and software. On the one hand, the hardware must have a high capability to allow its reconfiguration for different applications, but on the other hand, any particular application will only require a subset of those capabilities. Dynamic voltage and frequency scaling (DVFS) and gating techniques allow disabling un-needed capabilities to some extent, while heterogeneous systems can provide application-specific acceleration

http://dx.doi.org/10.5772/66678

facilities (see Section 3.2). Ultimately, the most energy-efficient implementation of an application is one that has dedicated hardware suited for the application and context, a solution, i.e., often economically infeasible. More generalized hardware leads to lower cost, but typically greater energy consumption.

The most significant factor in improving the energy efficiency of software is the developer. Developers can adjust their software to be more efficient on their chosen platform; however, this requires detailed understanding and creativity. It is currently impossible for automated processes to fully customize an algorithm to take full advantage of available hardware features, and thus, it is up to the creativity of the developer to do so. To create the best opportunities for efficiency, the developer should be provided with as much information about their software's energy consumption as possible: at all levels, from the switching cost to static power and system-level energy consumption. Enabling the developer in such a way will allow for energy-aware exploration of the design space, allowing informed decisions to be made so that the developer can pursue minimal energy. Such benefits are not limited to individual applications either: developers of software libraries, compiler optimizations, code schedulers, and any other software facility will be able to make energy-orientated design decisions. Better software design for hardware should also enable better selection of hardware. Understanding precisely how a piece of software consumes energy would allow more informed selection of heterogeneous systems, particularly if the heterogeneous hardware's features could be characterized in a way that can be matched against software features. Such a process would greatly simplify the benchmarking and evaluation required when selecting a platform to develop on.

In summary, new paradigms and tools for the design of software, based on energy transparency, are required if the energy consumption is to be reduced. Currently, the main drivers for software production are high performance, minimizing time for operation, and minimizing production cost. These drivers must include minimizing energy. For this to succeed, designers should be able to:

- Allocate compute resources in units of energy, not just time.
- Capture both execution duration and power efficiency.
- Force application developers to think about energy and understand and energy model of the software.
- Access energy-aware tools for developing energy-efficient software and code.

Tools such as energy modelling of software, software energy analysis methods, energy profiling, and metrics for numerics required for the development of energy-aware software and algorithms and of energy transparency are discussed in Chapters 5 and 6.

3.4. Communications

The ability to communicate information among devices, memory, storage, and systems is fundamental to ICT systems. While many concentrate on the energy from switching logic and memory devices, communication between devices and especially circuits or systems can be many orders of magnitude greater than logic operations. Efforts to reduce data transfer (and data reduction/compression methods) at the architecture and software level have been alluded in previous sections. Here, we summarize the different sources of the energy overhead of communications.

3.4.1. Chip-level: interconnects and electrical wires

Interconnect scaling has a significant impact on microarchitecture as it governs the speed at which data can be moved around a chip. In a micro-architecture with many cores, each with multiple functional blocks, the interconnect can become a bottleneck to performance. While transistors have benefited from performance increases as they have shrunk, interconnects have not, their delay product remaining close to constant [38]. The interconnect now plays a large part in the delays present in memory hierarchies. Thus, the structure of cache hierarchies—something, i.e., very much a micro-architectural choice—is governed not just by the speed of the storage cells, but the delays in moving data between those caches and the processor. Fast, first-level caches cannot be large, because it becomes impossible to transport data across them and still close timing constraints. Thus, modern high-performance processors have multiple cache layers, increasing in capacity and latency as they get further from the computational part of the processor.

For any electrical wire, the energy consumed to transmit a bit of information is related to the capacitance, C and the voltage, V. If the capacitance is approximated by $C \sim \varepsilon_0 L$ assuming an isolated wire where ε_0 is the permittivity of a vacuum, and L is the length, then the energy for transmitting a bit of information in a metal interconnect or wire can be approximated as

$$E_{\text{wire}} = \frac{CV^2}{2} \sim \frac{\varepsilon_0 L}{2}\left(\frac{k_B T}{e}\right)^2 \tag{1}$$

This gives a fundamental limit of 3×10^{-15} J per bit.m for isolated interconnects or wires [39]. The other fundamental limit is shot noise limit relating to the quantum of electric charge in electrons being transported along interconnects. At 300 K this corresponds to energy of 2.9×10^{-21} J/bit. In real systems, the actual values are significantly larger than the fundamental limits. In particular, the signal to noise and fan-out require significantly higher energy consumption especially when circuit speeds are high.

3.4.2. System-level: photonics and wireless

At the moment optical cables between boards and racks are available for Data Centres and HPC, but there is still significant copper interconnects before getting to the microprocessor, on-chip memory, or disk storage. The development of integrated Si photonics where the enormous yields of silicon foundries can be used to produce far cheaper and larger bandwidth (through parallel channels) has the potential to reduce energy per bit far faster than older technologies. Such technology is being developed for not only chip-to-chip photonic communications but also on-chip communications for the higher level and longer interconnects. If the volumes can be reached, then the costs should allow large-scale deployment. Miller [40] has investigated the requirements for chip-to-chip and on-chip photonic communications that provide an idea of the ultimate limits for photonic communications. The limits on individual photonic components is likely to be around 10 fJ/bit for the most power hungry, suggesting that the limit for short-distance photonic communications (<10 m distance) is around 100 fJ/bit.

For the rack-to-rack optical interconnects for exascale computing for 2019, the US Department of Energy calculated that every pJ/bit of optical power results in a total contribution of 0.8 MW for the complete system power [41]. As of 2014, a typical rack-to-rack optical interconnect at 40 Gbits/s is operating at around 40 pJ/bit. A number of authors have estimated the energy cost of sending information over the Internet [42]. For example in 2009, Baliga et al. [43] undertook a modelling study that included the energy consumption of the core, metro and edge, access, and video distribution networks. This included the energy consumption from switching and transmission equipment. They found energy consumption per bit of 75 μJ at low access rates that decreased from 2 to 4 μJ at an access rate of 100 MB/s. This study estimated that the Internet communication components alone accounted for 0.4% of electricity consumption in broadband-enabled countries at that point, with this percentage predicted to increase significantly as the bandwidth increases. They modelled how the energy per bit will reduce depending on how fast photonic technology improves and with a 10% rate of improved energy efficiency provides 700 nJ/bit by 2023, while a 20% rate of improvement results in 120 nJ/bit. The biggest issue is whether these reductions are aggressive enough to counteract the increase in the bandwidth of Internet and microwave wireless traffic. A key finding of the study was that the predicted rate of improvement with the future photonic technology in reducing the energy consumption per bit was not sufficient to reduce the energy consumption in the future as demand increases.

The IoT will require increased communication bandwidth, but as many of the systems are battery-powered and require wireless transmission, there are significant incentives to minimize communication bandwidth, distance, and time to save battery power. High-definition video is another issue. As use of high-definition video on demand increases, the systems architecture of the Internet to reduce long-haul delivery requires investigation with an aim to reduce energy consumption which is proportional to the bandwidth. For wireless communications the energy to transmit a bit of information is given by

$$E_{\text{wireless}} = N_{\text{photons}} E_{\text{photons}} \tag{2}$$

where N_{photons} is the number of photons given for uniformly radiating wireless as $N_{\text{photons}} \sim \frac{4\pi r^2}{\lambda^2}$ and E_{photon} is the photon energy given by $E_{\text{photons}} = h\nu = \frac{hc}{\lambda}$. The energy to transmit a bit of information is therefore given by [39]

$$E_{\text{wireless}} \sim \frac{4\pi r^2 hc}{\lambda} \tag{3}$$

Figure 10 presents the ultimate energy per bit that can be transmitted over 10 m distance by wireless using the single photon limit calculation and compares this to different wireless technologies. All the wireless technologies are around the 60—200 nJ range per bit, which is many orders of magnitude above the fundamental limits indicating that there is significant potential for improvements in the energy consumption of wireless communications. Chapter 9 of Vol. 1 presents an introduction to power consumption in wireless sensor networks including power consumption assessment via modelling and measurements. The evolution and state-of-the-art of wirelessly communicating networks of embedded computers and their energy efficiency are described in Chapter 7.

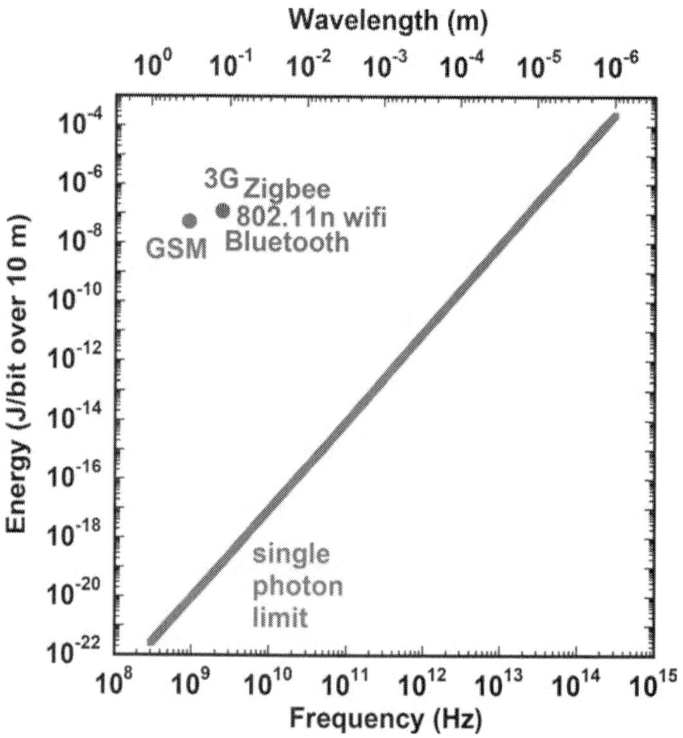

Figure 10. The energy per bit to transmit wireless data 10 m using a number of technologies at a range of transmission frequencies as compared to the fundamental single photon limit for omni-directional wireless transmission.

3.5. Energy sources and power management

Already many HPC cloud servers are being located so that renewable energy can be used to power at least part of the systems. Some of the best examples of low carbon renewables that can be used are photovoltaic, hydro, and wind (see **Figure 6**). In every case, the correct environment is required for each of the renewable technologies. A good example of this is photovoltaics that is being heavily used for a significant number of Data Centres for cloud computing. **Figure 11** provides the available energy averaged over 24 hours and 365 days of the year for a range of cities around Earth. The actual available energy over 24 hours will be these values times the conversion efficiency for the PV technology being used. The record PV efficiencies can be found at the charts published by the National Renewable Energy Laboratory (NREL) [http://www.nrel.gov/ncpv/images/efficiency_chart.jpg], but the majority of deployed PV solar farms use crystalline Si PV cells typically with starting efficiencies of 20–22%. Therefore, a solar farm in Athens will produce about 41 W/m^2 and to power a 10 MW HPC or cloud Data Centre will require at least 250,000 m^2 of PV area. Due to dust and dirt along with the harsh

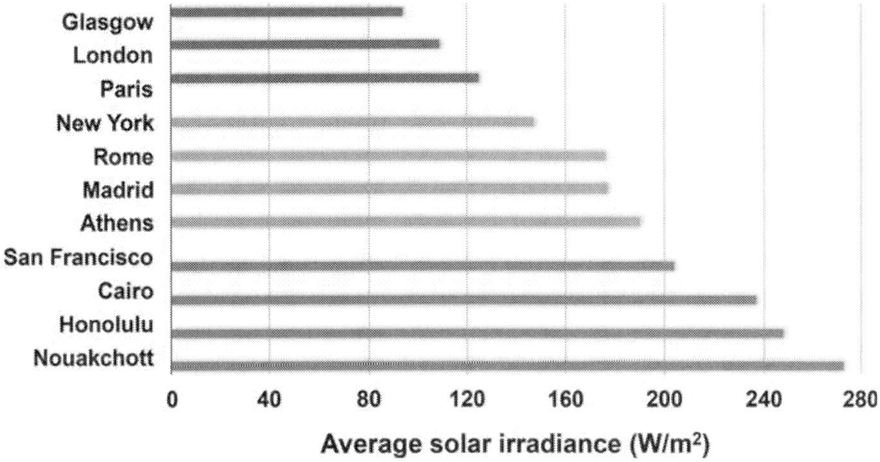

Figure 11. The average solar power available at different points on the earth integrated over 24 hours and 365 days of the year. The peak power is significantly higher than these values, but these are the ones available 24/7 (Source [44]).

environment that PV operates in, the efficiency drops off with time so a significantly larger area is required for long-term sustainable energy generation. Also, as mentioned at the end of Section 2, a significant challenge is that substantial energy storage is required to capture the energy that is only available during the day so it can be deployed at night.

The control and conversion of electric power through efficient power electronics is another issue to consider for power management on a large scale. While national grids that transmit electricity to industry and domestic properties for use all transmit using high-voltage AC to minimize losses, all ICT systems operate with DC power supplies requiring power transformation and management. At present switch-mode power supplies dominate AC to DC conversion for most ICT devices such as PCs, laptops, smart phones, and mobile phones. Such converters are used since they have greater efficiency than other technologies such as linear power supplies because the switching transistor dissipates little power when acting as a switch and spends very little time in any high-dissipation energy transition. Silicon power switches include power MOSFETs and insulated gate bipolar transistors (IGBT) depending on the power and regulation requirements. To improve efficiencies in power conversion, most research is concentrating on developing new materials for power switches which reduce the ohmic losses and the on-resistance. SiC and GaN are the main materials being developed as the wider bandgap and higher electrical conductivities should improve the conversion efficiencies. As an example, GaN is predicted to have 50% higher conversion efficiency than Si for power switches and just converting all electrical drives to GaN is predicted to save 9% of the electricity consumption in the UK that corresponds to removing the equivalent generation capacity of five advanced gas cooler nuclear reactors (UK Department for Business, Innovation and Skills October [45]). The potential for energy savings across Europe and the world are enormous and have been predicted to be of the order of €1400 Bn per annum (UK Department for Business, Innovation and Skills October [45]) if GaN technology fully replaces present Si power switches. An area of primary interest is complete integration of both modular and granular electronic power converters on-die and within package (power supply-on-chip) [46]. Following the trends of More Moore and More than Moore, this integration would deliver ever-greater current density, voltage regulation, and optimized control, form factor reduction, high efficiency, and cost reduction to meet performance requirements of emerging ICT systems.

Energy and power management in smart autonomous sensors needs to be embedded within the system architecture utilizing energy harvesting from the environment, energy storage devices, and efficient power distribution architecture. Similar to renewable energy, the energy sources must be chosen dependent on the operation environment. Energy harvesting mechanisms typically convert ambient kinetic energy, wasted energy (heat), and electromagnetic radiation (light, RF) into useful electricity and the challenge is to maximize the available extraction and conversion efficiency. Vibrational energy harvesters around the size of a drink are able to extract up to 5 mW of power from kinetic energy in pumps, trains, or vehicles. Such systems are now deployed in many industrial and transport applications. Thermoelectrics can be used to convert waste heat into electricity. Thermoelectric and photovoltaic (PV) energy conversion seems to be most promising for a large range of applications for small-to-medium power generation (and large for PV solar farms; see **Figure 6**). A temperature gradient is required to generate electricity with the voltage and power output dependent on the size of this temperature gradient. Such systems are used for smart thermostat controls of heaters and heating systems and also are being developed for industrial applications and automotive (to improve fuel consumption of vehicles). For autonomous sensors which can use PV, it is a reliable source during the day and batteries or super capacitors can be used to store up for nighttime use. Even in northern EU countries, the generation levels are around 20 W/m^2 per day, providing significant energy for most autonomous sensors. Indoor PV has significantly lower energy available typically 100 times lower than outdoor direct sunlight. Also most indoor available light is diffuse, scattered off different surfaces, and so capturing this light is far more difficult than capturing direct sunlight. The efficiency to capture diffused sunlight results in PV cells with efficiencies only up to about 10% at present. This is still useful for many autonomous sensors. A comprehensive presentation of energy harvesters can be found in Vol. 1. The physics of thermoelectrics and PV is presented in Chapter 9 for completeness.

As for HPC/Data Centres, energy storage and power management present a significant challenge for smart autonomous sensors too. There is a need to bridge the gap between the energy requirements for the ICT system operation and the energy supply from harvesting sources and storage to enable true autonomy (incl. wireless communication). In addition to decreasing the energy demand from the electronics and increasing the efficiency of the energy harvesters, an increase in storage capacity of batteries and the efficiency of power management electronics is required. The ideal scenario is an ICT system that integrates energy storage with energy harvesting components to provide power on demand to the electronic sensing, communication and display components, and it operates over the anticipated device lifetime of years. Ultimately, the energy supply components, harvesting and storage should occupy a footprint on chip no larger than the electronics it drives, with 1 mm^2 as an attractive long-term target for both. No such energy storage component exists today.

http://dx.doi.org/10.5772/66678

Batteries are the most common energy storage option, and since their introduction in the 1990s, lithium-ion batteries have exhibited the highest energy density that has been gradually improved in the intervening period from ~200 Wh/l to ~700 Wh/l through the use of improved materials and processing (see Chapter 6 of Vol 1 for a comprehensive overview of battery materials and architectures). Solid-state microbatteries that can be processed/integrated on silicon substrate exhibit similar volumetric energy densities with micron scale thin film materials. This necessitates a large area format consistent with the large areas or volumes required by solar or vibrational harvesting, respectively. Solid-state devices do offer capacity retention over thousands of cycles [47] that matches the device lifetime requirements. In the 2D, thin-film geometry, current deposition techniques, and lithium-ion diffusion characteristics in the solid-state limit the electrode thickness to several micrometers resulting in a battery dominated by the substrate and other inactive cell components. As thin films, these 2D formats typically exhibit energy densities of ~6 Wh/mm^2 or 0.2 J/mm^2. An energy budget of 1 mWh/day can support a wireless sensor node (WSN) used in building energy management with sensing and transmission every 20 minutes [46]. Clearly, significant advances are required for energy storage devices to meet the demand in a reasonable footprint. The key challenges are to realize improved energy storage in a significantly decreased footprint for ICT integration and high rate (power) capability during device interrogation and to decrease recharge time. These challenges require:

- Higher energy density materials, particularly at the cathode, where the current material, $LiCoO_2$, is 25 times less energy dense than the lithium metal anode.
- 3D or 1D active materials structuring with increased aspect ratio providing additional material (stored energy) with respect to planar commercial thin film microbatteries.
- Nanoscale active materials with improved electronic conductivity or core/shell structures [48] to facilitate high rate solid-state lithium ion transport.

Furthermore, current solid-state microbatteries cannot meet the needs of the ICT systems at peak power during measurement and transceiver operation and require a hybrid energy source with a supercapacitor. This is due to the lithium diffusion limitation in the solid-state electrolyte and cathode. On the other hand, the solid-state construction does facilitate the use of lithium metal anodes that have a large energy capacity (3600 mAh/g) in comparison with the typical carbon anodes (372 mAh/g) of most lithium-ion batteries. If nonsolid-state electrolytes are to be utilized for higher power outputs, then alternative high-energy intercalation anodes such as Sn (990 mAh/g), Ge (1600 mAh/g), or Si (4200 mAh/g) will be required to prevent dendritic short circuits on cycling. Core/shell [49] versions of these anodes may be required to alleviate mechanical stresses, leading to poor cycling behaviour and improve the electronic conductivity to access all of the high aspect ratio structures.

Disruptive battery technologies such as a Li/sulphur or Li/air can achieve the energy storage requirements of the ICT community. A theoretical energy density of 2.8 mWh/mm^3 (10 J/mm^3) has been estimated for a Li-air battery [50] with nonaqueous electrolytes. It is recognized that the Li/air and other high energy systems have to overcome many obstacles before they will be in widespread deployment, but many researchers predict that this could be over the next 10—15 years. A very challenging scaling requirement is shown in **Figure 12** and that must be coupled with decreased power requirements in the ultimate device. It is clear from the values in **Figure 12** that both new materials and 3D structuring of the materials are required to enable the decreased footprint desired for autonomous systems. An increasing focus on improved nanoarchitectures, electrode materials, and integrated current collectors is required to surmount these obstacles and deliver high-energy density solutions to meet the needs of the electronics industry. On the other hand, supercapacitors are emerging as an attractive candidate to complement advanced lithium batteries [52]. The key advantages of supercapacitors for such applications include high-power density, faster charge and discharge rates, and improved cycle-life.

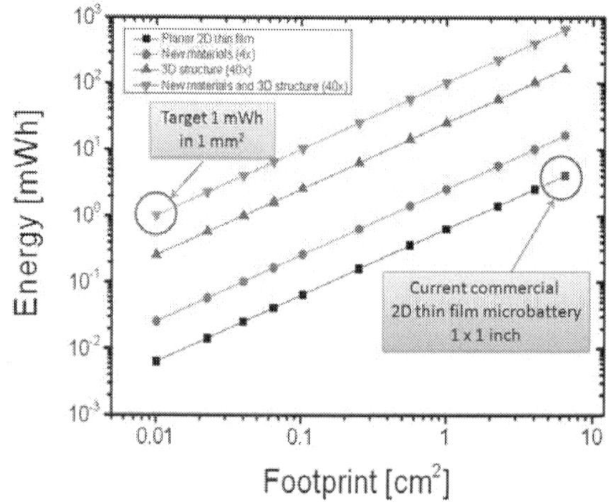

Figure 12. A roadmap for microbattery energy storage requiring the development and integration of new materials that could yield up to four times improvement in stored energy and micro or nanoarchitectures to increase the material quantity and surface area to deliver 1 mWh of energy in a 1 mm^2 footprint (Source [51]).

While traditional electrochemical supercapacitors can provide very high specific energy, preventing leakage of the supercapacitors liquid electrolyte solution in a portable or implantable device may be challenging. Also, since exposure of the liquid electrolyte solution to moisture in air can adversely affect its performance, special manufacturing needs are required that to date have prevented integration of supercapacitors into standard manufacturing processes. Thus, there is a market pull for a mechanically durable high-performance solid-state supercapacitor that can be fabricated using standard semiconductor manufacturing processes. While some solid-state supercapacitors (SSCs) with highly desirable power and energy density attributes have been demonstrated, significant practical challenges still remain in order to deliver

http://dx.doi.org/10.5772/66678

high-performance products to meet the future demands of ICT applications. Desired targets for future solid-state supercapacitors to meet these demands would include:

- High energy density >10 Wh/kg.[1]
- High power density >1 × 10^6 W/kg.
- Wide operating temperature window (−20°C to +70°C).
- Low equivalent series resistance (ESR) <10 mΩ.
- Long cycle life >1 × 10^6 cycles.

4. Conclusion

The key message that has become clear from investigating each of the levels of the system stack is that overall energy efficiency can be optimized more aggressively when the design at one level understands the energy issues at another level. This is especially true for levels that control the behaviour of another level, e.g., circuit architecture on devices or software on circuit architecture. At the start of ICT systems in the 1960s and 1970s, it was possible for engineers to understand all levels of the system stack and design accordingly. As each level has become more complex and the number of transistors and lines of code have moved from thousands into many billions, it has become more difficult for people to understand all levels of the system. A clear message is that only through the joint understanding of how different levels of the stack must be designed to reduce energy consumption will optimum ICT solutions that minimize the use of energy be found. As communication is the biggest energy consumption, and sustainable energy is environmental dependent, what is the optimum distribution and location of servers for the cloud if it is driven by sustainable energy ICT? Similarly, as the energy per bit of wireless communication scales with distance, what is the optimum distribution of smart autonomous sensors and data exchange to minimize energy consumption? It should be pointed out that in some circumstances minimizing communication is not the lowest energy solution—e.g., the processing required for heavy compression can cost more energy than you save under most transmission scenarios. If significant energy reductions are to be achieved, suitable education and tools are essential where an expert (e.g., in software) is provided with sufficient knowledge to understand the energy impact (of code) at the other levels of the stack. Realizing this vision requires interdisciplinary research at the boundaries of multiple scientific domains (materials science, physics, electrical engineering, software engineering, and mathematics), as well as developing and integrating innovations in several research areas, e.g., materials modelling and fabrication, device and computer engineering, cooling design, large-scale computing system simulation, software generation and optimization, statistical network modelling, and model predictive control theory.

[1] The energy density of current commercial off-the-shelf supercapacitors is relatively low at less than ~ 10 Wh/kg [53] with current solid state supercapacitor equivalents at ~ 1 Wh/kg [54].

Acknowledgements

This chapter has been broadly based on the Strategic Research Agenda (SRA) of the ICT-Energy project funded by the European Union (Ref: 611004). Many people have contributed in developing this SRA: Giovanni Ansaloni (EPFL), David Atienza (EPFL), Micheal Burke (Tyndall National Institute), Zbigniew Chamski (MpicoSys), Adrian Cristal (Barcelona Supercomputer Centre), Kerstin Eder (University of Bristol), Pablo Garcia (EPFL), Vincent Heuveline (Heidelberg University), Steve Kerrison (University of Bristol), Louise Krug (BT), Andrew Lord (BT), Jeremy Morse (University of Bristol), Simeon Oxizidis (International Energy Research Centre at Tyndall) James Rohan (Tyndall National Institute), Martin Wlotzka (Heidelberg University, M. Fernando Gonzalez Zalba (Hitachi Cambridge Laboratory), Olivier Zendra (Inria), Victor Zhirnov (Semiconductor Research Corporation). Contributions were also taken from the ICT Energy workshop "Future Energy" in the ICT Research Agenda on the 15 September 2015 in Bristol, where this SRA was presented, and a number of working groups provided feedback that has been included. Not all the names of the participants have been included in the above list. The participants included a wide range of academics through the system stack and companies including ARM and Intel.

Author details

Giorgos Fagas[1,*], John P. Gallagher[2], Luca Gammaitoni[3] and Douglas J. Paul[4]

*Address all correspondence to: georgios.fagas@tyndall.ie

1 Tyndall National Institute, UCC, Cork, Ireland

2 IMT, Roskilde University, Roskilde, Denmark

3 NiPS Laboratory, Department of Physics & Geology, University of Perugia, Perugia, Italy

4 School of Engineering, University of Glasgow, Glasgow, UK

References

[1] S.G. Anders Andrae and T. Edler, "On Global Electricity Usage of Communication Technology: Trends to 2030," Challenges 6, 117–157 (2015).

[2] United Nation's Brundtland Commission "Our Common Future" (1987)—Available at link: http://www.un-documents.net/our-common-future.pdf

[3] G. Fagas, L. Gammaitoni, D. Paul and G. Abadal Berini (eds.), ICT—Energy—Concepts Towards Zero—Power Information and Communication Technology," InTech ISBN 978-953-51-1218-1, DOI: 10.5772/55410 (2014)–Available at link: http://www.intechopen.com/books/ict-energy-concepts-towards-zero-power-information-and-communication-technology

[4] http://www.top500.org.

[5] R.H. Dennard et al., "Design of Ion Implanted MOSFET's with very Small Physical Dimensions," IEEE Journal of Solid-State Circuits 9(5), pp. 256–268 (1974).

[6] F. Pollack "New Microachitecture Challenges in the Coming Generations of CMOS Process Technology," 32nd International Symposium on Microarchitecture (Micro32) (1999)—Available at link: http://research.ac.upc.edu/HPCseminar/SEM9900/Pollack1.pdf

[7] International Atomic Energy Agency, "Nuclear Power Plant Design Characteristics" (2007)—Available at link: http://www-pub.iaea.org/mtcd/publications/pdf/te_1544_web.pdf

[8] ARM White Paper, "big.LITTLE Technology: The Future of Mobile Making very high performance available in a mobile envelope without sacrificing energy efficiency" (2013) —Available at link https://www.arm.com/files/pdf/big_LITTLE_Technology_the_Futue_of_Mobile.pdf

[9] http://www.gartner.com/technology/research.jsp

[10] United Nations Department of Economic and Social Affairs "World Population Prospects: The 2012 Revision" (2012)—Available at link: http://esa.un.org/wpp/unpp/panel_population.htm

[11] Cisco White Paper "The Zettabyte Era: Trends and Analysis" (2015)—Available at link: http://www.cisco.com/c/en/us/solutions/collateral/service-provider/visual-networking-index-vni/VNI_Hyperconnectivity_WP.html

[12] M. Pickavet, W. Vereecken, S. Demeyer, P. Audenaert, B. Vermeulen, C. Develder, D. Colle, B. Dhoedt and P. Demeester, "Worldwide Energy Needs for ICT: The Rise of Power-Aware Networking," 2nd International Conference on Advanced Networks and Telecommunication Systems, 2008, pp. 1–3 (2008).

[13] W. Van Heddeghem, S. Lambert, B. Lannoo, D. Colle, M. Pickavet and P. Demeester, "Trends in Worldwide ICT Electricity Consumption from 2007 to 2012," Computer Communications 50, pp 64–76 (2014).

[14] J. Malmodin, A. Moberg, D. Lundén, G. Finnveden and N. Lövehagen, "Greenhouse Gas Emissions and Operational Electricity Use in the ICT and Entertainment & Media Sectors," Journal of Industrial Ecology 14, pp. 770–790 (2010).

[15] Global e-Sustainability Initiative "SMARTer2020: The Role of ICT in Driving a Sustainable Future" (2013)—Available at link: http://gesi.org/portfolio/report/72

[16] International Energy Authority, "Key World Energy Statistics" (2014)—Available at link: http://www.iea.org/statistics/

[17] R. Landauer, "Irreversibility and Heat Generation in the Computing Process," IBM Journal of Research Development 5, 183–191 (1961).

[18] V.V. Zhirnov, R.K. Cavin III, J.A. Hutchby and G.I. Bourianoff, "Limits to Binary Logic Switch Scaling—A Gedanken Model," Proceedings of the IEEE 91, pp. 1934–1939 (2003).

[19] International Technology Roadmap for Semiconductors—Available at link: http://www.itrs.net/

[20] D.E. Nikonov and I.A. Young, "Uniform Methodology for Benchmarking beyond-CMOS Logic Devices," Proceedings IEDM 12, pp. 576–672 (2012).

[21] M. López-Suárez, I. Neri and L. Gammaitoni, "Sub-kBT Micro Electromechanical Irreversible Logic Gate," Nature Communications 7, 12068 (2016).

[22] B.K.G. Brill, "The Invisible Crisis in the Data Center: The Economic Meltdown of Moore's Law," White Paper, Uptime Institute, Rev. 2, 2007–12—Available at link: http://www.mm4m.net/library/The_Invisible_Crisis_in_the_Data_Center.pdf

[23] H. Mamaghanian, N. Khaled, D. Atienza Alonso and P. Vandergheynst, "Design and Exploration of Low-Power Analog to Information Conversion Based on Compressed Sensing," IEEE Journal of Emerging and Selected Topics in Circuits and Systems 2(3), pp. 493–501, (2012).

[24] S. Benatti, B. Milosevic, F. Casamassima, P. Schonle, P. Bunjaku, S. Fateh, Q. Huang, L. Benini, "EMG-based Hand Gesture Recognition With Flexible Analog Front End," in Proceedings of International IEEE Conference on Biomedical Circuit and Systems (BIOCAS 2014), Lausanne, Oct 2014

[25] N. Planes et al., "28 nm FDSOI Technology Platform for High-Speed Low-Voltage Digital Applications," Proceedings of Symposium VLSI (2012).

[26] J.-P. Colinge and J. Greer, "Nanowire Transistors," Cambridge University Press (Cambridge, UK), ISBN: 978-1107052406 (2016).

[27] A. Sridhar, M.M. Sabry, P. Ruch, D. Atienza, B. Michel, "PowerCool: Simulation of Integrated Microfluidic Power Generation in Bright Silicon MPSoCs," oral, ICCAD, November 2014, San Jose, CA, USA.

[28] N. Rasmussen, "Guidelines for Specification of Data Center Power Density," (2005)—Available at link: http://www.apcdistributors.com/white-papers/Cooling/WP%20120%20Guidelines%20for%20Specification%20of%20Data%20Center%20Power%20Density.pdf

[29] V. Hanumaiah and S. Vrudhula, "Energy-efficient Operation of Multicore Processors by DVFS, Task Migration and Active Cooling," IEEE Transactions on Computers 63(2), pp. 349–360, (2012).

[30] P. Bassett and M. Saint-Laurent, "Energy Efficient Design Techniques for a Digital Signal Processor," 2012 IEEE International Conference on IC Design Technology (ICICDT), pp. 1–4 (2012).

[31] ARM Cortex-A15 MPCore Processor Technical Reference Manual, ARM, pp. 53–63 (2013).

[32] R. Bahar and S. Manne, "Power and Energy Reduction via Pipeline Balancing," Proceedings 28th Annual International Symposium on Computer Architecture, 2001, pp. 218–229 (2001).

[33] H. Zeng, J. Wang, G. Zhang, and W. Hu, "An Interconnect-Aware Power Efficient Cache Coherence Protocol for CMPs," IEEE International Symposium on Parallel and Distributed Processing, IPDPS 2008 pp. 1–11 (2008).

[34] A.Z. Jooya and M. Analoui, "Program Phase Detection in Heterogeneous Multi-Core Processors," 14th International CSI Computer Conference, CSICC 2009 pp. 219–224 (2009).

[35] K. Changkyu, S. Sethumadhavan, M. S. Govindan, N. Ranganathan, D. Gulati, D. Burger, and S. Keckler, "Composable Lightweight Processors," 40th Annual IEEE/ACM International Symposium on Microarchitecture, MICRO 2007 pp. 381–394 (2007).

[36] Z. Rakossy, T. Naphade, and A. Chattopadhyay, "Design and Analysis of Layered Coarse-Grained Reconfigurable Architecture," 2012 International Conference on Reconfigurable Computing and FPGAs (ReConFig) 2012 pp. 1–6 (2012).

[37] K. Roy and M.C. Johnson, "Software Design for Low Power," in "Low Power Design in Deep Submicron Electronics," W. Nebel and J. Mermet (Eds.), Kluwer Nato Advanced Science Institutes Series, Vol. 337. Kluwer Academic Publishers, Norwell, MA, USA, pp 433–460 (1997).

[38] M. Bohr, "A 30 Year Retrospective on Dennard's MOSFET Scaling Paper," IEEE Solid-State Circuits Society, Newsletter 12(1), pp. 11–13 (2007).

[39] V.V. Zhirnov, "Fundamentals of Energy Consumption in ICT Devices," NIPS/ICT Energy Summer School (2014)–Available at link: http://www.nipslab.org/sites/nipslab.org/files/NiPS2014_Zhirnov_en%20cons%20ICT.pdf

[40] D.A.B. Miller, "Device Requirements for Optical Interconnects to Silicon Chip," Proceedings of IEEE 97(7), pp. 1166–1185 (2009).

[41] http://www.hoti.org/hoti20/slides/Fuad_Doany_IBM.pdf

[42] J. Baliga et al., "Green Cloud Computing: Balancing Energy in Processing, Storage and Transport," Proceedings of IEEE 99(1), pp. 149–167 (2011).

[43] J. Baliga et al., "Energy Consumption in Optical IP Networks," Journal of Lightwave Technology 27(13), pp. 2391–2403 (2009).

[44] G. Boyle, "Renewable Energy: Power for a Sustainable Future," Oxford University Press (Oxford, UK), ISBN: 978-0199545339. (2012).

[45] UK Department for Business, Innovation and Skills, "Power Electronics: A Strategy for Success," October (2011)–Available at link: https://www.gov.uk/government/uploads/system/uploads/attachment_data/file/31795/11-1073-power-electronics-strategy-for-success.pdf

[46] C.Ó. Mathúna, T. O'Donnell, R. Martinez, J.F. Rohan and B. O'Flynn, "Energy Scavenging for Long-Term Deployable Mote Networks," Talanta 75, 613–623 (2008).

[47] N.J. Dudney, "Solid-state thin-film rechargeable batteries ", Journal of Materials Science and Engineering B-Solid State Materials for Advanced Technology 2005, 116, 245.

[48] M. Hasan, T. Chowdhury, J.F. Rohan, "Nanotubes of Core/Shell Cu/Cu2O as Anode Materials for Li-ion Rechargeable Batteries ", Journal of the Electrochemical Society A682, 157(2010).

[49] L.F. Cui, R. Ruffo, C.K. Chan, H.L. Peng, Y. Cui, "Crystalline-Amorphous Core-Shell Silicon Nanowires for High Capacity and High Current Battery Electrodes", Nano Letters 9, 491 (2009).

[50] J.P. Zheng, R.Y. Liang, M. Hendrickson, E.J. Plichta, "Theoretical energy density of Li–air batteries", Journal of the Electrochemical Society 155, A432 (2008).

[51] W. Wang, J.F. Rohan, N. Wang, M. Hayes, A. Romani, E. Macrelli, M. Dini, M. Filippi, M. Tartagni and D. Flandre, Chapter 9—"Smart Energy Management and Conversion in Beyond CMOS Nanodevices," 1 (2014) Editor F. Balestra, Wiley, pp 249–276, ISBN: 978-1-84821-654-9.

[52] F. Gonzalez and P. Harrop, "Batteries & Supercapacitors in Consumer Electronics 2013–2023: Forecasts, Opportunities, Innovation," IDTechEx (Cambridge, UK) (2014).

[53] A. Burke, Z. Liu, H. Zhao, "Present and future applications of supercapacitors in electric and hybrid vehicles," IEEE International Electric Vehicle Conference, DOI: 10.1109/IEVC.2014.7056094 (2014)

[54] P. Banerjee, I. Perez, L. Henn-Lecordier, S. B. Lee, and G.W. Rublof, "Nanotubular metal-insulator-metal capacitor arrays for energy storage" Nature Nanotechnology 4, 292 (2009)

Chapter 2

Fundamentals on Energy in ICT

Luca Gammaitoni

Additional information is available at the end of the chapter

http://dx.doi.org/10.5772/66973

Abstract

This chapter deals with the fundamental physical aspects of the use of energy in ICT devices. Here we discuss questions such as "what is the theoretical minimum energy required to process information?", "what is the minimum energy required to transmit information from one point to another?" and "are these limits practically reachable and under what conditions?" While dealing with these relevant questions, we are mostly concerned with providing to the reader a clear and intuitive understanding of what is going on and what the underlying physics aspects are.

Keywords: energy, information, noise, efficiency, entropy

1. Introduction

This chapter deals with the fundamental physical aspects of the use of energy in ICT devices. Here, we discuss questions like *what is the theoretical minimum energy required to process information? What is the minimum energy required to transmit information from one point to another?* And *are these limits practically reachable and under what conditions?*

Most importantly, in dealing with these relevant questions, we will be mostly concerned with providing to the reader a clear and intuitive understanding of what is going on and what are the underlying physical aspects, more than showing rigorous mathematical demonstrations. In fact, these can be found in many university textbooks (some listed at the end of the chapter) and missing some rigor will hopefully not harm the validity of the reasoning.

Dealing with fundamentals in ICT necessarily implies dealing with physics. In fact any ICT device, being it a complex microprocessor with billions of transistors interconnected or a simple binary logic gate is, first of all, a physical system. As such its functioning is subjected to the laws of physics. Regarding the implications of the use of energy in such devices we are thus referred to the very elegant theory of thermodynamics. In this theory, many scientists through the years have accumulated all the knowledge developed in dealing with energy and its transformations. Thanks to the work of scientists like Sadi Carnot, Emile Clapeyron, Rudolf Clausius and William Thomson (Lord Kelvin), studies on how energy could be used with profit in machines invented earlier by Thomas Newcomen and James Watt to transform heat into work, brought us the notion of entropy and the second law of thermodynamics that put limits on the efficiency of such machines.

Steam engines from the dawn of the industrial revolution are not much different from nowadays ICT systems if you look at them by a merely physics point of view. In both classes of devices, we are dealing with the transformation of energy from heat to work and from work to heat. Surprisingly, 200 years have passed after the work of Carnot but we are still intrigued by the problem of defining the efficiency of these transformations even if, at difference with the past, today the object of our quest has moved from the heat engines of the industrial revolution to the tiny devices of modern ICT.

It is a common statement that future ICT will be characterized by nanoscale devices that will process information while dissipating significant amount of energy, i.e. while transforming work into heat. In this perspective it seems natural to consider an ICT device as a novel info-thermal machine: it inputs information and energy (in the form of work) and outputs information and energy (in the form of heat).

In the following, we will discuss in detail these aspects trying to make clear what are the underlying fundamental physical laws that govern the use of energy in ICT devices. We will proceed as follows:

2. What is information processing and how this can be done with machines?

3. Basics on thermodynamics laws.

4. Digital computing and the physics of switches.

5. Energy efficiency, Landauer reset and reversible computing.

6. Energy bounds on communication as information transfer processes.

2. What is information processing and how this can be done with machines?

In this section, we discuss the fundamentals in information processing. We introduce the notion of 'amount of information' and its digital representation. Most importantly, we discuss how a physical system can be used to do information processing and how this has to do with the laws of physics and with energy transformation processes in particular.

Let us start with a fundamental question: what does it mean 'information processing'? Before we can answer this question, we need to introduce the notion of 'information'.

This notion was introduced, in the framework that is of interest here, by Claude Shannon (1916–2001) in 1948 in his attempt to formulate a 'mathematical theory of communication'.

http://dx.doi.org/10.5772/66973

During a communication process, there is a message that needs to be transmitted from one point to another. In this perspective, the 'amount of information' is a quantity that can be associated with a given message.

To illustrate this concept, let us assume that we want to transmit a text message. Something like 'Hello my friend, what's up?'. This message is composed of a number letters and punctuation symbols. Let us suppose that this message is part of a much longer message so that we can assign to each symbol a given probability to be part of this message. Typically, if the message is written in a given language, we can use the frequency of each letter in that language as the probability. If we call a generic symbol x_i (this can be a letter like 'a' or 'A' or a punctuation symbol like ';') then we can indicate the probability to find it in our message is $p(x_i)$. At this point, we can define the amount of information that is carried by each symbol x_i (according to Claude Shannon definition) as the number H_i given by:

$$H_i = -K\, p(x_i) \log p(x_i) \tag{1}$$

where K is simply a constant and 'log' represents the logarithmic function. By the moment that the probability $p(x_i)$ is a number between 0 and 1, the log $p(x_i)$ is a negative number and thus the resulting H_i is a positive quantity. H_i is also sometimes called 'entropy', in analogy with the physics quantity as introduced by Gibbs previously (see below).

When we want to transmit a message, it is more practical to transmit only a small number of different symbols in order to avoid possible confusion between two similar symbols. In fact, you can easily realize that it is easier to transmit the Latin alphabet, with say 25 symbols, than the Japanese katakana with 48 different characters. Thinking about it, people realized that the most confusion-avoiding (noise-resistant) way to do this is to associate each symbol (character) you want to transmit to a number and then to represent the number in base 2, with only two different digits, e.g. '0' and '1'.

In synthesis, when we transmit a message we transmit a stream of symbols '0' and '1' that can be associated with numbers that are associated with letters and punctuations. Traditionally, these symbols '0' and '1' are called *bit* as a contraction of the words '*b*inary dig*it*'.

In most common long messages, the probability to find a '0' or a '1' is the same and is thus $p(0) = p(1) = 0.5$. According to this, the amount of information transported by a message composed of n bits with n_0 symbols '0' and n_1 symbols '1' (and $n = n_0 + n_1$) is:

$$H = -K\left(n_0 \frac{1}{2} \log \frac{1}{2} + n_1 \frac{1}{2} \log \frac{1}{2}\right) = K \frac{1}{2} (n_0 + n_1) \log 2 \tag{2}$$

If we chose $K = 2$ and assume that log 2 = 1 (this is true if we admit the base 2 for the logarithmic function), we have

$$H = (n_0 + n_1) = n \tag{3}$$

Thus, the amount of information in a message coded in bits is equal to the number of bits in the message.

Within this framework, 'information processing' is what we do when we manipulate (= do operations on) the bits of a message. By the moment, bits are numbers, 'information processing' is substantially equivalent to computing.

So we are back to another fundamental question: how do we do computing? This question may seem a bit naïve by the moment that we are all used to deal with the act of computing since when we were kids. Clearly computing is associated with dealing with quantities, represented by numbers: how to count them, how to add, subtract and in general how to transform numbers. This is absolutely correct. However, here we would like to focus on the fact that, at its very fundamental roots, the act of computing can be associated with very simple physical manipulations like moving a ball from one pot to another or changing the position of a pebble in a row or a column. This was known since the old times: the word 'calculus' (form which 'calculations') comes from Latin and designated small stones that Romans used to account for quantities, i.e. to perform computations.

Manipulating physical objects is thus at the base of computing and we have shown that such manipulation has the power of transforming not only numbers but more generally any kind of symbol, as it is normally carried out in modern ICT devices that process information, like when we read an e-mail or change an image on a screen.

By the moment that information processing/computing can be associated with the change of bits, in order to perform this activity we need two very important components:

1. a physical system capable of assuming two different physical states
2. a way to induce state changes in this physical system (typically a force).

We are not going to spend time dealing with the quite philosophical definition of physical system by intending with it any object, device or phenomena that can be studied by physics. The notion of physical state is slightly more delicate. With it we mean a set of measurable quantities whose value can be used to distinguish unambiguously two different outcomes, as an example shown in **Figure 1**.

Here we have

1. The physical system, made by a pebble and two bowls. The two states are represented by the measurable quantity 'position of the pebble': state '0' = pebble in left bowl; state '1' = pebble in the right bowl;
2. The way to induce state changes represented by a force that brings around the pebble.

According to this example, we can perform information processing activity simply by changing the position of the pebble, according to certain rules, with the underlining idea that, while we do these changes, we are at the same time changing the value of the symbol '0' and '1' associated with the system state. Devices that obey the rules (1) and (2) are called *binary switches*.

In modern computers, binary switches are made of transistors. These are electronic devices (**Figure 2**, left) that satisfy the two required conditions:

http://dx.doi.org/10.5772/66973

1. The two states are represented by the measurable quantity 'electric voltage' at point V_{OUT}. As an example state, '0' = $V_{OUT} < V_T$; state '1' = $V_{OUT} > V_T$; with V_T a given reference voltage;
2. The way to induce state changes represented by an electromotive force applied at point V_{IN}.

Figure 1. Two bowls and one pebble are sufficient to realize a binary switch.

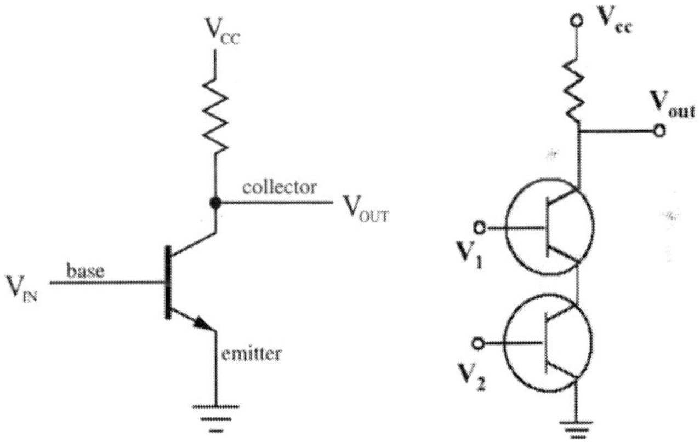

Figure 2. Left: transistor. Right: a combination of two transistors to realize a NAND gate.

By combining binary switches, we can perform all the information processing operations required. As an example, we mention the NAND gate (a universal logic gate) that can be realized by interconnecting two transistors (see **Figure 2**, right).

Now, the question that we want to address is the following: what is the minimum energy required to process information? In order to answer this question, we have to briefly recall the basic laws of one of the most elegant physics theories: thermodynamics.

3. Basics on thermodynamics laws

Thermodynamics is the theory that deals with concepts like energy, work, heat, entropy and their use in physical systems. In this section, we present in a concise way the fundamental laws of thermodynamics [1]. It will help us to understand what can we do and what we cannot do with energy.

The fundamental laws were considered through a period of approximately 100 years during which wrong assumptions, brilliant experiments and hard work characterized the work of a bunch of great scientists. Among them we list Thomas Newcomen (1664–1729) who built the first practical steam engine aimed at pumping water out of coal mines and James Watt (1736–1819) who soon after realized an improved version of the same machine. The laws of thermodynamics were considered to provide understanding and tools to the engine makers. This effort was carried out in few decades by some remarkable scientist: Émile Clapeyron (1799–1864), Sadi Carnot (1796–1832), Rudolf Clausius (1822–1888), and William Thomson (Lord Kelvin) (1824–1907).

The laws of thermodynamics do not tell us much about what energy is but they are very good in ruling what can we do and what we cannot do when we change the energy content of a body by exchanging heat and work.

The first law of thermodynamics is about the *conservation of energy*. It states that the total energy of a physical system remains the same during any transformation the system can go through, provided we take into account how much work the system does and how much heat the system exchanges (i.e. work and heat balances out).

It was first proposed by Julius Robert von Mayer (1814–1878) and subsequently reviewed by James Prescot Joule (1818–1889) and Hermann Ludwig Ferdinand von Helmholtz (1821–1894). Conservation of energy is strongly believed to be true and, to some extent, it is a self-sustaining law: it is so strongly believed that in every instance we observe a possible violation, we think harder to discover some way in which energy could have been overlooked and if we cannot find a way, well… we invent a new kind of energy. In past we did so at the beginning of 1900 when Albert Einstein introduced the mass-energy equivalence, to account for the 'missing mass' during a nuclear transformation.

The second law is about how much energy in the form of heat we can draw from a system in order to do work. Specifically, the second law shows that there are limitations to the amount of work we can get from a given amount of energy present in the form of heat. There are few equivalent formulations of these laws. We list here the two most popular, ascribed to Rudolf Clausius and Lord Kelvin:

Clausius: 'No process is possible whose sole result is the transfer of heat from a body of lower temperature to a body of higher temperature'.

Kelvin: 'No process is possible in which the sole result is the absorption of heat from a reservoir and its complete conversion into work'.

An important consequence of the second law, discovered by Sadi Carnot in 1824 (when he was 28 years old), is that there is a limit to the efficiency of a thermal machine. In the publication entitled *Réflexions sur la Puissance Motrice du Feu* ('Reflections on the Motive Power of Fire') Carnot generalized the concept, popular at that time, of 'steam engine' by introducing the novel concept of 'thermal machine'. A thermal machine is a physical system that can exchange heat and work with its surroundings. Carnot showed that the efficiency of any thermal machine operating between two temperatures is bounded by a quantity that is a function of the two temperatures only and does not depend on the features of the machine (nor the material, nor the geometry, nor the functioning principles). It was a great result, indeed.

Soon after the work of Sadì Carnot, Rudolf Clausius used his result to introduce a new physical quantity that is useful in describing exactly how much heat can be changed into work during the transformation. He suggested the name 'entropy' for this quantity.

The reasoning behind the introduction of this quantity, the entropy, is the following: to operate a thermal machine, it is necessary to find a cyclic transformation during which heat is changed into work. The cycle is necessary because you want to operate the machine continuously and not just once: every time that a cycle is completed you get some work. By reiterating the cycle you can get any amount of work you need. First of all, Clausius proved a theorem that states that during a cyclic transformation, if you do the transformation carefully enough not to lose any energy in other ways (like friction), then the algebraic sum of the heat exchanged with the external (considered positive the heat that goes into the system and negative the heat that leaves the system) divided by the temperature at which the exchanges occur is zero:

$$\oint \frac{dQ}{T} = 0 \tag{4}$$

An important aspect is that the cycle does not depend on the specific path that you take. Moreover, being a cycle, you start and end at the same state. Clausius concluded that, from the previous integral, it does exist a state function S defined as

$$S_B - S_A = \int_A^B \frac{dQ}{T} \tag{5}$$

(or in differential form $dS = dQ/T$). The function S is a state function (it only depends on the state) and represents a novel physical quantity called entropy.

In addition, Clausius showed that if during the transformation, you are not careful enough and you lose energy (due to friction), than the inequality holds instead of the equality:

$$\oint \frac{dQ}{T} \leq 0 \tag{6}$$

The transformation like this is also called an *irreversible transformation*. It is easy to show that if we take and irreversible transformation to compute the entropy, we end up with under-estimating the change:

$$S_B - S_A \geq \int_{A\,\mathrm{irr}}^{B} \frac{dQ}{T} \tag{7}$$

It is important to point out that in practical cases it is practically unavoidable to have some kind of friction, thus, the inequality holds. In the very special case, in which we have transformation where we do not have any heat exchange (sometimes called adiabatic transformation), then the right hand of the inequality is zero and the final entropy is always larger than the initial one.

The concept of irreversible transformation is a bit tricky. You can have an irreversible transformation even if there is no apparent friction. This is the case, for example, of the so-called free expansion of a gas. James Prescot Joule in 1845 has shown with a remarkable experiment that you can have a gas to expand freely (without doing any work) from a smaller container to a larger one, without any heat exchange. In this case, the irreversibility of the transformation comes from the fact that during the free expansion, the gas is out of equilibrium, i.e. the usual thermodynamic quantities like temperature, pressure and volume are not well defined due to the fact that the gas is expanding and while parts of the gas are still at a certain temperature (with a given mean velocity), other parts of the gas may show different mean velocities.

If we consider an infinitesimal transformation we have:

$$dS \geq \frac{dQ}{T} \text{ or } TdS \geq dQ \tag{8}$$

where the equal sign holds during a reversible transformation only. The previous equation is often considered the formulation of the second law of thermodynamics [1]. By putting in contact, a physical system that is at temperature $T1$ with a heat reservoir that is at temperature $T2 > T1$, then some heat will be transferred from the reservoir to the system. Accordingly, the integral is positive and the entropy of the system increases (meaning that this process can occur without any work). The other way around phenomenon by which heat is transferred from the system to the reservoir does not occur because it would require a decrease of entropy (second principle) and thus we conclude that during a spontaneous transformation (i.e. without external work) the entropy always increases. We can make the entropy of our system decrease (e.g. like in a refrigerator) but we have to add work from outside [1].

Another way of looking at this formulation of the second principle is the following. In the general case of irreversible transformation, instead of using the inequality, we can write the previous expression as:

$$TdS = dQ + E_d \tag{9}$$

where E_d is the additional energy dissipated during the transformation, meaning that, when we want the entropy to decrease a quantity TdS, we need to spend an amount of minimum energy equal to dQ. If we cannot do things carefully enough to reach a reversible (i.e. lossless, where the quantity E_d is always zero) transformation condition, then we need to spend $dQ + E_d$. Instead, if we want to do the transformation where the entropy increases, then we do not need to spend any minimum amount of energy. Entropy increase can come for free!

Back to the Clausius inequality, it is useful to interpret the quantity TdS in a reversible transformation as the amount of heat (meaning thermal energy) that cannot be used to produce work [1]. In other words, during the transformation, even if we are carefully enough not to waste energy in other ways, we cannot use all the energy that we have to do useful work, part of this energy will go into the entropy change. If we are not carefully enough the situation is even worst and we get even less work. The limitation in heat transformation is usually quantified by the introduction of the so-called *free energy*. The concept of free energy was proposed by Helmholtz in the form: $F = U - TS$. The free energy F quantifies the maximum amount of energy that we can use to do useful work, when we have available the internal energy U of a system with entropy S.

The introduction of entropy was aimed at quantifying the limitations on the use of heat to produce work. However, it is not an exaggeration to say that entropy, in general, remained for many years an obscure quantity, whose physical sense was difficult to grasp. It was the work of Ludwing Boltzmann (1844–1906) that shed some light on the microscopic interpretation of the second law (and thus the entropy). Boltzmann proposed an interpretation of the second law, i.e. the natural tendency of systems to evolve (via spontaneous transformations) towards the state characterized by the increased entropy as the tendency of a system to attain an equilibrium condition identified as the most probable state, among all the states the system can be in.

In the idealized world considered by Boltzmann, physical systems are gasses made by many small parts represented by colliding little spheres. Let us consider an ideal gas made by N such particles in the form of tiny hard spheres of mass m that can collide elastically (thus conserving kinetic energy and momentum). Let us suppose that these particles are contained in a box with one of the walls consisting in a moving set of mass $M = Nm$. The set is connected to a spring of elastic constant k, as shown in **Figure 3**, and is at rest [1].

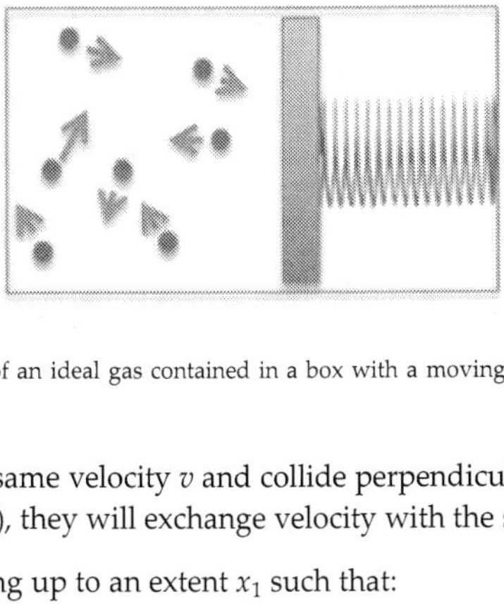

Figure 3. Pictorial representation of an ideal gas contained in a box with a moving set connected to a spring. The gas particles move randomly.

If all the particles have the same velocity v and collide perpendicularly with the moving set at the same time (see **Figure 4**), they will exchange velocity with the set.

This will compress the spring up to an extent x_1 such that:

$$\frac{1}{2}M\,v^2 = \frac{1}{2}k\,x_1^2 = U \tag{10}$$

By a purely mechanical point of view, this is a mere transformation of kinetic energy into potential energy. We can always recover the potential energy U when we desire and use it to perform work. The work will be exactly U. In this case, we can completely transform the kinetic energy of the gas particle into work. How comes? Well, in this case we are considering

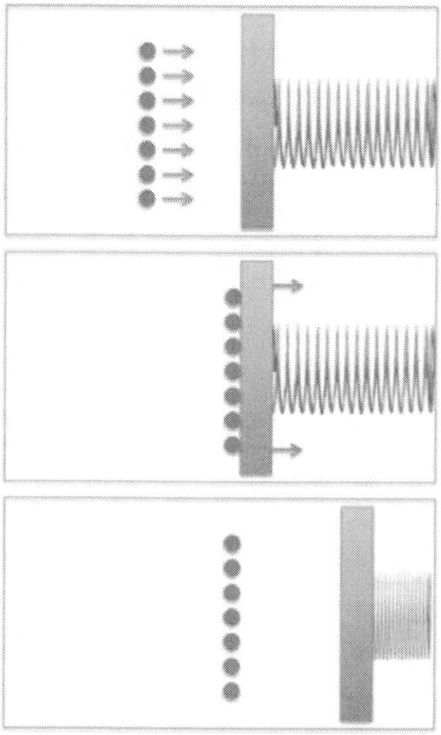

Figure 4. Pictorial representation of the ideal gas. Top: all the particles move in ordered way from left to right with parallel velocities. Middle: the particles hit the piston and exchange velocity with it. Bottom: The piston compresses the spring and reach zero velocity. All the energy is now stored as potential energy.

a very special configuration of our gas (unique indeed) where all the particles are moving accordingly in parallel lines. If we put randomly the particles in the box, what is on the contrary the most probable configuration for their arrangement? Based on our experience (and on some common sense as well), the most probable configuration is one where all the particles are moving with random direction (but same velocity) in the box. The kinetic energy of the gas is still the same (so is its temperature T) but in this case, the movable set will be subjected at random motion with an average compression of the spring such that its average energy is U/N. This is also the maximum work that we can recover from the potential energy of the movable set. Thus, it appears clear that, although the total energy U is the same in the two cases, in the second case we have no hope of using the greatest part of this energy to perform useful work. As we have determined, when we introduced the definition of free energy, the quantity that limits our capability of performing work is the entropy. Following this definition, the system that has the smaller entropy has the larger capability of performing work. Accordingly, we can use the entropy to put a label on the useful energetic content of a system. Two systems may have the same energy but the system that has the lower entropy will have the 'most useful' energy.

http://dx.doi.org/10.5772/66973

This example helped us to understand how energy and entropy are connected to the microscopic properties of the physical systems. In the simple case of an ideal gas, the system energy is nothing else than the sum of all the kinetic energies of the single particles. We can say that the energy is associated with 'how much' the particles move. On the other hand, we have seen that there is also a 'quality' of the motion of the particles that is relevant for the entropy. We can say that the entropy is associated with 'the way' the particles move. This concept of 'way of moving' was made clear by Boltzmann at the end of 1800, who proposed for the entropy the following definition:

$$S = k_B \log W \tag{11}$$

where k_B is the famous Boltzmann constant and W is called the 'number of configurations' and represents the number of ways we can arrange all the particles in the system without changing its macroscopic properties. In the previous example, we have only one way to arrange the N particles so that they are all parallel, aligned and with the same velocity while we have a very large number of ways of arranging the N particles to be a randomly oriented set of particles with velocity v. Thus, it is clear that in the second case, the value of the entropy is much larger than that in the first case (where it is indeed zero).

The Boltzmann formula refers to a case where all the microstates are equiprobable. The extension to the more general case with microstates with different probabilities was proposed by Josiah Willard Gibbs (1839–1903):

$$S = -k_B \sum_i p_i \log p_i \tag{12}$$

where p_i represents the probability of the microstate i.

We have seen above that during a spontaneous transformation, the entropy of the system increases. This can occurred without any change in the energy of the system itself as it was shown by Joule in the famous experiment of free gas expansion. Let us consider our previous example where all the particles move along parallel lines. Let us suppose that the trajectories are not perfectly aligned. Initially nothing happens but, due to a small misalignment, sooner or later a collision between the particles can happen and, collision after collision, the entire group of particles evolves into a randomly moving group. This is clearly a spontaneous transformation. By the moment that the collisions are elastic the energy of the system has not changed but the system entropy has rapidly increased from the zero initial value up to its maximum value. Conversely, the free energy has reached its minimum value. It is interesting to ask: can we bring the system back to its initial condition? The answer is yes but in order to do it we need to spend some energy as required by the second principle. How much? Clearly we need to spend a minimum of $T\Delta S$ of energy, where ΔS represents the difference in entropy between the final and the initial states. The bad news is that if we spend this energy and decrease the entropy back to its original condition, the energy that we spend does not change the total kinetic energy of the system that remains the same. However, having reduced the system entropy we have increased the Free energy and this improves our capability of extracting work from the system itself.

4. Digital computing and the physics of switches

In order to answer our initial question (*what is the minimum energy required to process information?*) we need to apply the thermodynamics concepts that we have just learned to the binary switches that are the basic elements of any information processing device.

As we have discussed in Section 2, a binary switch is a physical system that obeys the two rules (1) and (2) that here we restate as follows:

1. a physical system capable of assuming two different physical states: S_0 and S_1
2. a set of forces that induce state changes in this physical system: F_{01} produces the change $S_0 \rightarrow S_1$ and F_{10} produces the change $S_1 \rightarrow S_0$.

If we think about it, we can easily realize that there exist at least two classes of devices that can satisfy these rules. We call them *combinational* and *sequential* devices [1].

Combinational devices are characterized by the following behaviour: when no external force is present, under equilibrium conditions, they are in the state S_0. When an external force F_{01} is present, they switch to the state S_1 and remain in that state as long as the force is present. Once the force is removed they come back to the state S_0. Popular examples are represented by relays (**Figure 5**) and also by transistors, today widely exploited in modern computing devices to make logic gates. A combinational device is a network of combinational switches.

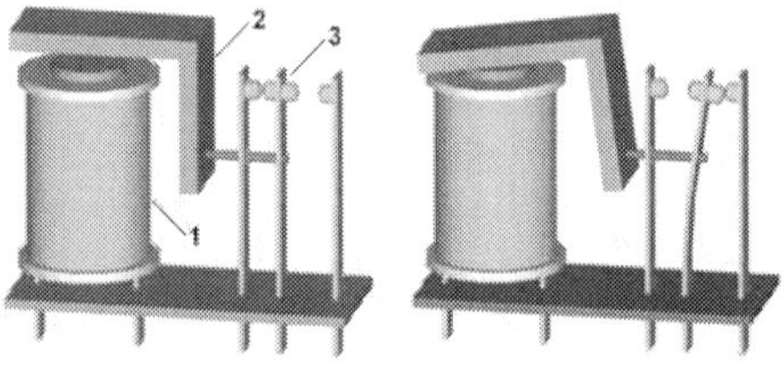

Figure 5. Electro-mechanical switch (relay) as a combinational device: when an external force (magnetic induction force) is applied the switch changes its state from open to close) and goes back to the initial state (open) when the force is removed. Picture obtained from Wikipedia: https://commons.wikimedia.org/wiki/File:Relay_principle_horizontal.jpg.

Sequential devices are characterized by the following behaviour: if they are in the state S_0, they can be changed into the state S_1 by applying an external force F_{01}. Once they are in the state S_1 they remain in this state when the force is removed. The transition from state S_1 to S_0 is obtained by applying a new force F_{10}. In contrast to the combinational device, the sequential device remembers its state after the removal of the force. This memory lasts for a time that is short compared to the system relaxation time. In fact, if one waits long enough, the sequential device relaxes to equilibrium that, in a symmetric binary switch, is characterized by a 50% probability to be in the S_0 state and 50% probability to be in the S_1 state. This relaxation process is unavoidable in any real physical system that is operated at finite temperature. However, in all practical cases the relaxation time is usually much longer than any operational time; hence, the sequential device can be considered a system that remembers the last transition. Examples

http://dx.doi.org/10.5772/66973

include electronic flip-flop and DRAM (dynamic random access memory): the complex 'storage capacitor + transistor'. They are employed in computers to perform the role of registers and memory cells. A simple mechanical example of sequential binary switch is the switch illustrated in **Figure 6**.

Figure 6. Mechanical binary switch as a sequential device: when an external force is removed it stays in the last state for any interval of time, shorter than the relaxation time.

In order to discuss the energetic behaviour of the two classes of binary switches, we need to introduce a dynamical model that is capable of representing the action of the force and the switch mechanism. In order to do so, we use a simple model based on a single degree of freedom $x(t)$ that represents the system state (this can be the position of a pebble or the value of some electric voltage or current, as we discussed above). This quantity $x(t)$ must be subjected to constrains and forces that make it to behave according the two rules (1) and (2) and also some time evolution equation, according to physics.

For both classes of devices, we can use the following equation:

$$m\ddot{x} = -\frac{d}{dx}U(x) - m\gamma\dot{x} + \xi(t) + F \tag{13}$$

where m represents the inertia of our system, F is an external force that can be applied when we want to change state, and γ is the frictional force that represent dissipative effects in the switch dynamics and

$$U(x) = \frac{1}{2}x^2 \quad \text{for combinational devices} \tag{14}$$

$$U(x) = -\frac{1}{2}x^2 + \frac{1}{4}x^4 \quad \text{for sequential devices} \tag{15}$$

In general, $U(x)$ is a potential function that has the role of confining $x(t)$ in a well-defined region. What is $\xi(t)$? This is a stochastic force and represents the role of fluctuations that are unavoidably present due to a finite temperature. These fluctuations are responsible for the relaxation process that we discussed above. In a macroscopic binary switch, this term is quite small compared to the other terms in the equation of motion and is usually neglected. However, when we deal with micro- to nanoscale devices, like in modern binary switches, its role might be relevant and its presence cannot be neglected [2].

The fluctuating force $\xi(t)$ is represented here by a zero average stochastic process that is defined in statistical terms. Due to its presence, the equation of motion is a stochastic equation and its solution is usually described in statistical terms. $P(x, t)dx$ represents the probability for the quantity x to be at time t within the interval between x and $x + dx$ and is a relevant quantity to describe the system dynamics.

Here, we can define the two distinguishable physical states S_0 and S_1, as follows: the state S_0 is realized when $x < x_{TH}$; the state S_1 is realized when $x > x_{TH}$, and x_{TH} is a value of the quantity x that can be chosen conveniently. Due to the presence of fluctuations, the two physical states are assumed with a certain probability given by:

$$p_0(t) = \int_{-\infty}^{x_{TH}} P(x,t)dx \text{ and } p_1(t) = \int_{x_{TH}}^{+\infty} P(x,t)dx \tag{16}$$

The switch event in a combinational device is illustrated in **Figure 7**.

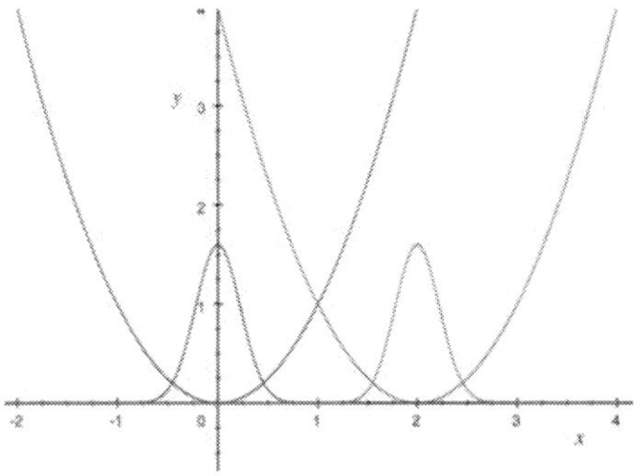

Figure 7. Switch event for the combinational device: the application of a constant force, $F_{01} = -F_0$, produces a net displacement of the $p(x)$ from the one in black (centered around $x = 0$) and corresponding to the state S_0 to the one in blue (centered around $x = 2$) and corresponding to the state S_1.

Here, the application of a constant force, $F_{01} = -F_0$, produces a net displacement of the $p(x)$. By setting properly the value of the threshold, we can easily realize the switch from S_0 to S_1. According to the combinational character of our device, once the force is removed, the system reverts back to the initial state S_0. In order to compute the amount of energy required for this switch, we should take into account the work done by the forces acting on the system. The stochastic force does not do (on average) any work because it is a zero mean force. The dissipative force does a negative work that is proportional to the switch speed. The external force acts on the potential and is a conservative force, thus through an entire cycle its work is null.

The role of the dissipative force and of the fluctuating force can be, more properly, discussed within the thermodynamics framework that we have previously introduced. In fact, their presence accounts for the existence of a coupling of our quantity $x(t)$ with a thermal bath that is responsible at the same time for the fluctuating part of the dynamics (i.e. the random force $\xi(t)$) and the dissipative part (i.e. the damping constant γ). Indeed the two are connected through a famous relation called the fluctuation-dissipation theorem [1] established by Harry

Theodor Nyquist (1889–1976) in 1928, and demonstrated by Callen and Welton in 1951. This relation is:

$$G_R = \frac{mK_BT}{\pi}\gamma \tag{17}$$

where G_R represents the intensity of the fluctuation with white noise spectrum [1].

Due to the existence of the thermal bath, thermodynamics sets the rule for the energy balance during the switch process. Specifically, if we conduct the switch process from the initial state S_0 to the final state S_1 we need to spend a minimum energy (i.e. producing a heat Q that goes into the thermal bath). In general, we have

$$TdS = dQ + E_d \tag{18}$$

If the transformation is carried out in a reversible manner, $E_d = 0$ and the amount of dissipated heat is, according to Clausius, $dQ = TdS$. In a cyclic operation, this is clearly zero because T is constant and S is a function of state only. On the other hand, if the transformation is not reversible, the amount of dissipated energy is E_d and is larger than zero.

Could this be realized in practice? In a recent experiment, Lopez-Suarez and et al. [3] has built a micro-cantilever that can be operated as a combinational device. They showed that by slowing down the switching operation, E_d can be made arbitrarily small, thus confirming that the minimum energy required to process information with a combinational device is indeed zero.

Let us now consider the switch event in a *sequential device*.

For this case [1], the definition of the switch event itself must be reconsidered. Previously, the switch event was defined as the change from an equilibrium position (e.g. at rest at the bottom of the potential well) to another equilibrium position (e.g. at rest at the bottom of the displaced potential well). In this bistable potential, however, the particle is never at rest at the bottom of a single well: due to the presence of the fluctuating force, the particle will be randomly oscillating around the potential minima, with occasional jumps between the two wells. Since the potential is symmetrical and we have a zero-mean fluctuating force, the two states S_0 and S_1 have the same probability. This implies that the probability density distribution at equilibrium $P(x,t) = P(x)$ is stationary and symmetric, as represented in **Figure 8**.

When the particle is initially at rest at the bottom of the left well, after some time τ_1 it starts to oscillate around the potential minima and after some longer time τ_2 it will jump into the right well and eventually back into the left well and so on. The time τ_1 and τ_2 are random variables. Their mean values $t_1 = <\tau_1>$ and $t_2 = <\tau_2>$ (with $t_2 > t_1$) can be computed on the bases of the features of the potential $U(x)$ and the stochastic force $\xi(t)$. They are usually addressed as the *intra-well* relaxation time and the *inter-well* relaxation time and, in general, they represent, respectively, the average time the system takes to establish equilibrium within one well and the average time it takes to go to global equilibrium. Since t_2 depends exponentially on the barrier height between the two wells, in practical switches the barrier height is chosen to be large enough to guarantee that $t_2 >> t_1$.

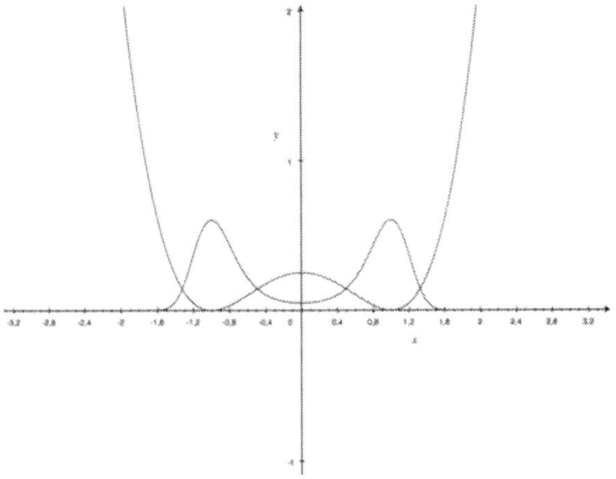

Figure 8. Bistable potential $U(x)$ with superimposed the probability distribution $P(x,t) = P(x)$ at equilibrium.

Based on these considerations, we can define the switch event as the transition from an initial condition towards a final condition, where the initial condition is defined as $\langle x \rangle < 0$ and the final condition is defined as $\langle x \rangle > 0$. With the initial condition characterized by:

$$p_0(t) = \int_{-\infty}^{0} P(x,t)dx \cong 1 \text{ and } p_1(t) = \int_{0}^{+\infty} P(x,t)dx \cong 0 \tag{19}$$

and the final condition by

$$p_0(t) = \int_{-\infty}^{0} P(x,t)dx \cong 0 \text{ and } p_1(t) = \int_{0}^{+\infty} P(x,t)dx \cong 1 \tag{20}$$

The conditions are reversed for a switch event from S_1 to S_0.

In order to produce the switch event, we proceed as follows: set the initial position at any value $x < 0$ and wait a time t_a, with $t_1 << t_a << t_2$, then apply an external force F for an elapsed time t_b to produce a change in the $\langle x \rangle$ value from $\langle x \rangle < 0$ to $\langle x \rangle > 0$. Then remove the force. In practice, it will be necessary to wait a time t_a after the force removal in order to verify that the switch event has occurred, i.e. that $\langle x \rangle > 0$. The total time spent has to satisfy the condition $2\,t_a + t_b << t_2$.

Now that a switch event has been defined in this new framework, we can return to the question: what is the minimum energy required to produce a switch event?

As before, for the combinational device, it is quite easy to see that in order to minimize the energy dissipation, the role of the friction has to be negligible. In addition to this condition, we need to make sure that during the transformation no irreversible increase of the entropy takes place. The most common case (to be avoided) is the free expansion. During a free expansion, the system does not do any work and the entropy increases without energy expenditure. However, when we need to bring back the system to its original state we cannot perform the reverse operation without energy expenditure because we need to decrease the entropy and this cannot be performed for free. This condition is particularly relevant for a procedure that is often followed in the switch event. The procedure is shown in **Figure 9** and consists in five subsequent steps.

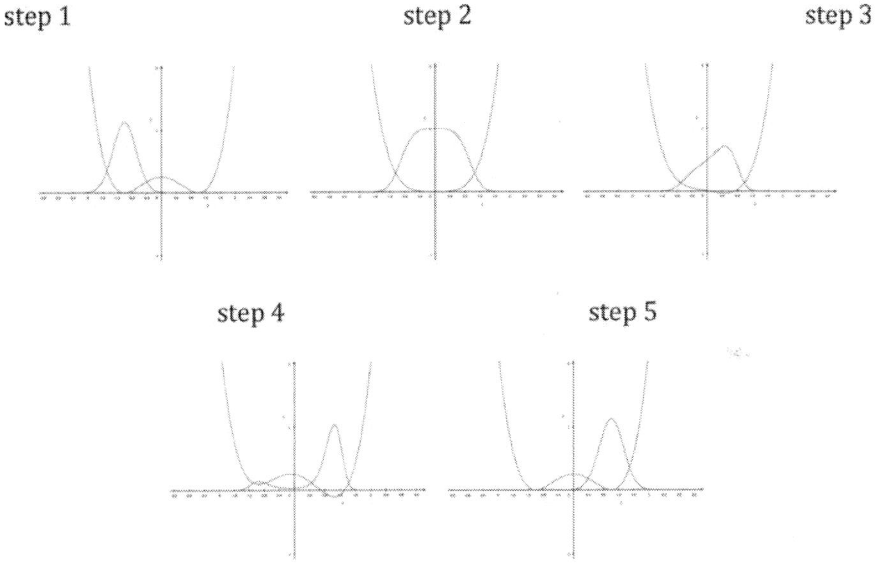

Figure 9. Potential $U(x) + F$. Equilibrium $P(x)$ for the different cases. Second procedure: Step 1 and Step 5, $F = 0$; step 2 $F = -x$; Step 3 $F = -x - F_1$; Step 4 $F = -F_1$.

We point out that from step 1 to step 2, the entropy of the system increases. During this transformation, the potential changes by lowering the barrier. At this point, the particle dynamics relaxes (in a very short time) to the new configuration and the entropy increases like in a free expansion. This is apparent by the change in the probability distribution and can be demonstrated by simply assuming that in step 1 we have $p_0 = 1$ and $p_1 = 0$, this gives $S_1 = -k_B \ln 1 = 0$. In step 2, $p_0 = p_1 = ½$ and thus $S_2 = -k_B (½ \ln ½ + ½ \ln ½) = k_B \ln 2$. Thus, $\Delta S = k_B \ln 2 > 0$. On the other hand, when there is a transition from step 2 to step 5, the entropy is reduced from S_2 to $S_5 = S_1 = 0$, thus $\Delta S = -k_B \ln 2 < 0$. According to the thermodynamics, these last steps cannot be performed without providing energy to the system and thus the minimum energy in this case is not zero [1].

Summarising, the conditions required to perform a switching event that takes zero energy, are: (1) the total work performed on the system by the external force has to be zero. (2) The switch event has to proceed with a speed arbitrarily small in order to have arbitrarily small losses due to friction. (3) No free expansion entropy increase during the procedure.

In the following, we show a procedure (called *zero-power* protocol in the literature [4]) that satisfies these three conditions. In order to satisfy condition (1), we apply a force that maintains the average position of the particle always close to the minimum of the potential well. In this case, the force is zero and thus the work is zero. In order to satisfy condition (2), we apply very slowly a change in the force. Finally, in order to satisfy condition (3), i.e. the probability density in state 0 and in state 1 is the same; apply a force that does not change the probability density along the path (constant entropy transformation). This can be done by applying a force that alters the potential, as shown in **Figure 10**. Such a procedure clearly satisfies the three conditions that we enunciated above.

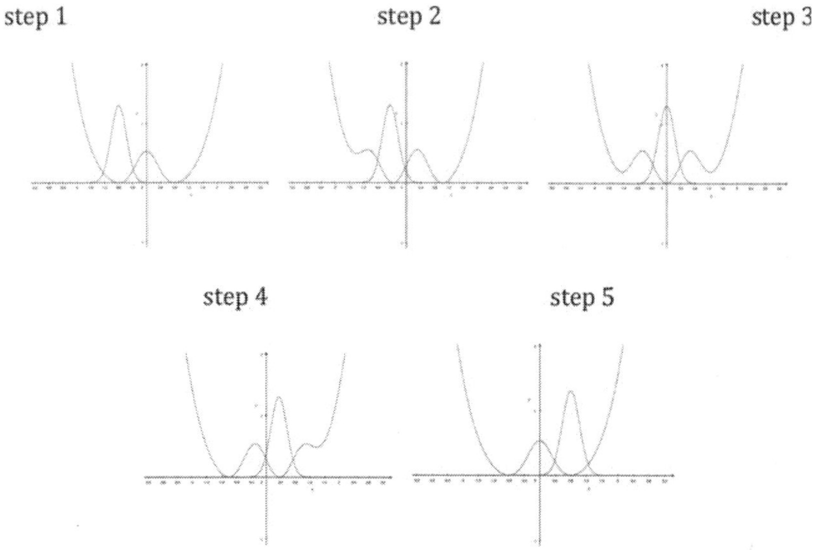

Figure 10. Potential $U(x) + F$. Equilibrium $P(x)$ for the different cases. Third procedure.

According to the reasoning developed in this procedure, we can conclude that also in the case of sequential devices, the minimum energy required to process information, i.e. to operate a switch, is indeed zero.

5. Energy efficiency, Landauer reset and reversible computing

In the previous sections, we have seen that a generic computing device can be considered as a machine that processes information while transforming work into heat. Pioneering research developed by J. Von Neumann and by R. Landauer in the last century has shown that information processing is intimately related to energy management ('information is physical' [5]). As a matter of fact, an ICT device is a machine that inputs information and energy (under the form of work), processes both and outputs information and energy (under the form of heat).

http://dx.doi.org/10.5772/66973

According to this model, energy efficiency during computation can be defined in terms of the quantity of input energy that is used for computation against the quantity that is transformed into heat. Thus, we can define energy efficiency as:

$$\eta = \frac{L-Q}{L} \tag{21}$$

L is the input energy (in the form of work) and Q is the wasted heat produced during computation. Clearly, it varies between 0 (a totally inefficient device: all the input energy is wasted into heat) and 1 (maximum efficiency where all the energy is used to perform computation and none is wasted into heat).

Based on this definition, it is clear that the effort to reach the maximum efficiency is equivalent to the effort to reach the zero heat produced condition $Q = 0$. We note that, in principle, a computing device can be operated by keeping, during computation, the total change in the internal energy $U = L - Q = 0$. So the minimum energy required is $L = Q$.

The question that we aim to address since the beginning of the chapter is the following: is there a limit to how small can we make Q?

This topic has been widely discussed in the scientific community since the beginning of the modern computers era. Based on our previous discussion, our question, by the moment that all the combinational and sequential devices can be made by interconnecting respective binary switches, translates into:

1. What is the minimum amount of energy required to operate a combinational switch?
2. What is the minimum amount of energy required to operate a sequential switch?

Based on the reasoning developed in the previous paragraph, we can now summarize the answer.

The answer to question (1) is $Q = 0$, provided that the switch operation is performed slowly enough in order to make negligible all the dissipative phenomena such as viscous damping and internal friction. Sometime this way of operating switches is called *adiabatic computing*.

The answer to question (2) is $Q = 0$, provided that the switch operation is performed slowly enough and that there is no irreversible entropy increase during the process (and the necessary and costly subsequent entropy decrease).

While the adiabatic computing condition is common to both classes of switches, the sequential devices demand an additional condition that requires some further discussion.

In fact, there exists a situation where the second condition in (2) cannot be satisfied. It is when the sequential switch is relaxed to equilibrium and the initial state condition is shared with equal probability by S_0 and S_1. In this case, it is common to say that 'the knowledge of the system state is lost'. In order to apply the *zero-power* protocol it is necessary to put the system into a known state with 100% probability. This operation is called *Landauer reset* and cannot be performed without entropy reduction [5]. By the moment that on average, it requires a reduction of the number of initial states from 2 to 1, the entropy decreases for a quantity $\Delta S = K_B$ Log 2 (it halves the state space) and this is necessarily associated with an amount of minimum energy to be dissipated $Q = K_B\, T$ Log 2.

In conclusion, we have shown that a computing device can be operated with arbitrarily low energy expenditure, provided that no *Landauer reset* is required. Otherwise, a minimum energy expenditure has to be accounted in the measure of $K_B\, T$ Log 2 per reset.

Needless to say that this conclusion regards what we have called the 'fundamental limit' in energy consumption, during computation. Clearly, other limits [1] can arise when we deal with practical realization of computing devices. However, even if these limits are presently much larger and more important for practical applications, we should not forget that they are associated with the specific technology used. By changing technology, we can (in principle) always aim at reaching the fundamental limits.

6. Energy bounds on communication as information transfer processes

In this section, we discuss the concept of information transmission and its implication on the amount of energy required. This topic has been addressed since the beginning of last century and has been put in the modern form by the father of information theory, Claude Shannon.

The starting point of Shannon's reasoning is the following: if we want to transmit a certain amount of information (i.e. a message) through a given communication channel (it could be air, vacuum, copper wire, etc.) we want that this information reaches its destination uncorrupted. The cause of potential corruption is called noise. Shannon was able to demonstrate that if in a given channel characterized by a bandwidth B, at the destination you can measure an amount of noise power N and a signal power S, then the maximum amount of information per unit time (bit per second) you can transmit (without corruption) is:

$$C = B \log_2\left(1 + \frac{S}{N}\right) \tag{22}$$

This relation is often addressed as the Shannon-Hartley theorem. We are interested in finding the minimum energy E_b that is required to transmit a single bit through a channel with a certain amount of noise. In order to find E_b, we need to express it in terms of the quantities in the Shannon's relation. By definition, the energy per bit is equal to the signal power S (energy per unit time) divided per capacity C (bit per second): $E_b = S/C$. On the other hand, the noise power N is equal to the noise spectral density N_0 (noise power per unitary bandwidth) times the bandwidth B: $N = N_0\, B$.

Thus, the previous relation becomes:

$$C_B = \log_2\left(1 + \frac{E_b}{N_0} C_B\right) \tag{23}$$

where we have introduced the quantity $C_B = C/B$, capacity per unitary bandwidth. The quantity E_b is readily obtained as:

$$E_b = \frac{(2^{C_B}-1)}{C_B} N_0 \tag{24}$$

If the channel bandwidth is much larger than the capacity (meaning that we transfer bits very slowly), we can take the limit when C_B goes to zero. By the moment that

$$\lim_{x\to 0} \frac{(a^x-1)}{x} = \ln a \tag{25}$$

we have $E_b = N_0 \ln 2$. The role of noise here is quite evident. How large it could be? Well, in principle, we can try to reduce the noise as much as possible but, even if we are able to suppress every other noise source, there is one source that cannot be avoided, which is the thermal noise, it is present in every physical system that is at finite temperature T and represent the natural oscillations of their elementary components (atoms and molecules). For the thermal noise, we have $N_0 = K_B\, T$. Thus, the minimum energy required for sending a bit of information, due to the fundamental noise limit, is

$$E_b = k_B T \ln 2 \tag{26}$$

As before, we would like to stress that this is fundamental limit that sets the minimum energy required. In practical systems, there are other noise sources that can play a relevant role as well.

One final word should be spent to compare this fundamental limit, i.e. the minimum energy required to transmit one bit, with the minimum energy required to do a switch event, i.e. the elementary step in the computation process. As we have seen that there is a minimum energy to be spent only in the case in which a Landauer reset is required. This amount of energy is dissipated in heat and it is definitively lost during information processing. For the minimum energy required to send one bit, instead, E_b is the energy associated with the physical signal that transmits the information. This quantity is not automatically dissipated in heat and, in principle, could be restored once the signal is received at the destination.

Author details

Luca Gammaitoni

Address all correspondence to: luca.gammaitoni@nipslab.org

NiPS Laboratory, University of Perugia, Perugia, Italy

References

[1] Gammaitoni L. Energy management at the nanoscale. (In ICT - Energy - Concepts Towards Zero - Power Information and Communication Technology, G. Fagas (Ed.), InTech (2014). DOI: 10.5772/57345. Available from: http://www.intechopen.com/books/

ict-energy-concepts-towards-zero-power-information-and-communication-technology/energy-management-at-the-nanoscale.

[2] See the chapter Zhirnov V, Cavin R, Gammaitoni L. Minimum energy of computing, fundamental considerations. In ICT - Energy - Concepts Towards Zero - Power Information and Communication Technology, InTech [1].

[3] Lopez-Suarez M, Neri I, Gammaitoni L. Sub k_BT micro-electromechanical irreversible logic gate, *Nature Communications* 7,12068 (2016) doi:10.1038/ncomms12068.

[4] Gammaitoni L, Chiuchiú, Madami M, Carlotti G. Towards zero-power ICT. Nanotechnology, 2015 Jun 5; 26(22):222001.

[5] Landauer R. Irreversibility and heat generation in the computing process, IBM Journal of Research and Development 1961; (5):183–191.

Interesting textbooks:

Information Theory: A Tutorial Introduction, James V Stone, Paperback – February 1, 2015.

On-line textbook: Information Theory, Inference, and Learning Algorithms, by David MacKay (2005). (available here: http://www.inference.phy.cam.ac.uk/mackay/itila/)

Chapter 3

Measuring Energy

Steve Kerrison, Markus Buschhoff,

Jose Nunez-Yanez and Kerstin Eder

Additional information is available at the end of the chapter

http://dx.doi.org/10.5772/65989

Abstract

This chapter provides an introduction to quantifying the energy consumed by software. It is written for computer scientists, software engineers, embedded system developers and programmers who want to understand how to measure the energy consumed by the code they write in order to optimize for energy efficiency. We start with an overview of the electrical foundations of energy measurement and show how these are applied by reviewing the most commonly found energy sensing techniques. This is followed by a brief discussion of the signal processing required to obtain energy consumption data from sensing. We then present two energy measurement systems that are based on sensing techniques. Both can be used to directly measure the energy consumed by software running on embedded systems without the need to modify the hardware. As an alternative, regression-based techniques can be used to infer energy consumption based on monitoring events during program execution using counters monitors offered by the hardware. We introduce the foundations of regression analysis and illustrate how an energy model for an ARM processor can be built using linear regression. In the conclusion, we offer a wider discussion on what should be considered when selecting an energy measurement technique.

Keywords: energy measurement, power, energy sensing, energy measurement systems, regression analysis

1. Introduction

Energy is now the limiting factor in electronic system design. To develop more energy efficient systems, energy must be taken into account during system design at all levels of abstraction. Low-power design has been a focus for hardware developers for several decades with impressive results in terms of low-power processors from embedded to high-performance systems.

However, beyond the hardware layer in the system stack there are further savings to be made. Experts expect these to be significantly higher than what the hardware can achieve on its own. For example, while dedicated low-power hardware can realize savings of 20% in multimedia processing, three to five times more could be saved with software support [1]. While energy efficient software development is only just emerging, it is clearly an essential part towards achieving energy efficiency of whole systems.

To support energy efficient software engineering, the energy consumed by software needs to be determined. Energy measurement is one way to achieve this, and this chapter provides the reader with an insight into techniques that can be used to measure the energy consumed by software running on embedded systems. The measurements can then be used to gain deeper understanding of how algorithms, languages, coding styles, data structures and compilers impact on the energy consumed during program execution; they also help engineers identify energy bugs in the software. Furthermore, energy consumption monitoring at runtime becomes feasible, and this enables dynamic adaptations to be introduced into systems to adjust their energy consumption in response to high or low levels of activity, or to accommodate varying levels of demand. Energy measurements also allow the creation of energy models. These can be used to estimate energy consumption either at design time or at runtime, without the need for measurements to be taken, thus saving the effort involved in setting up and measuring, often at the expense of accuracy, although acceptable error margins can be achieved with state-of-the-art modeling techniques.

Broadly speaking, energy consumption of electronic systems can be measured either directly or indirectly. The direct approach relies on sensing, which may require some instrumentation of the target hardware, often involving invasive procedures such as soldering, to install the components required for measurements to be taken. The knowledge and skills needed to accomplish this step are typically not within the repertoire of software developers, which can be a considerable barrier in practice. External energy measurement systems already come with sensing components built in and only require connection to the measurement points on the target hardware. This chapter introduces the fundamental concepts of direct energy measurement in Section 2 and presents two external measurement systems in Section 3.

The indirect approach infers energy consumption from other events that can be measured during program execution [2]. Modern architectures offer counters integrated into the hardware to collect statistics on the operation of the processor and memory system. Examples include the counters in the performance monitoring units on the Intel Xeon Phi [3] or the ARM Cortex A9 [4]. A variety of different events can be counted at runtime, such as the number of read or write misses at level 1 in the data cache or the number of data cache-dependent stall cycles in a pipeline. Regression analysis is applied to establish a correlation between these events and dynamic power dissipation. If this is successful, a predictive model can be built. Section 4 provides an introduction to regression analysis, which is illustrated with a worked example for the ARM Cortex A9 in Section 5.

http://dx.doi.org/10.5772/65989

Beyond direct measurement and inferring energy consumption indirectly through regression analysis, energy consumption information can also be obtained based on simulating a hardware design. This requires availability of the gate-level design description and layout information. A power estimation tool can then be used to obtain power data based on the amount of switching recorded during hardware simulation. This, together with performance data, provides energy consumption information. However, we will not cover this approach in more detail in this chapter, as we focus on energy measurement of real systems rather than designs.

2. Basics of direct measurement techniques

This section provides the necessary background to understand the processes by which a device's energy consumption can be directly sensed. The underlying physical principles defining electrical energy consumption are explained in detail, followed by a discussion of methods for observing and recording measurements.

2.1. Fundamental concepts

In the following sections we will briefly introduce some fundamental concepts and terminology of electrical engineering in the domain of energy measurement.

In general, measuring energy directly is infeasible, because energy is a virtual concept that quantifies the influence that things can have on their physical environment. Instead of measuring energy directly, we have to measure these physical effects and to deduce the energy that was involved in realizing these effects. The most important effect of energy in the realm of electricity is the ability to transport electrical charge, to build up electric and magnetic fields and to produce heat, light or other forms of radiation.

Electrical energy can be defined as the conversion of an electrical power, *P*, measured in watts, W, for a given amount of time. Since power changes continuously over a period of time $[t_0, t]$, energy is the integral of all converted power during this interval:

$$E(t) = \int_{t_0}^{t} P(t)\mathrm{d}t \tag{1}$$

Consequently, energy can be expressed in units of watt-seconds, Ws, which is equivalent to the unit joule, J, where 1 J = 1 Ws.

Electrical power is defined as the strength of an electric current, I, measured in amperes, A, caused by an electric "driving pressure", the potential difference, measured in volts, V. For a given point in time, the electrical power is defined as:

$$P(t) = V(t) \cdot I(t) \tag{2}$$

For certain classes of observed systems, assumptions can be made about voltage and current. For example, for many computational systems, the supply voltage is constant over a great period of time. This reduces the problem of measuring energy to the measurement of current and time. Thus, a device's energy consumption can be calculated as:

$$E(t) = V_{const} \int_{t_0}^{t} I(t) \mathrm{d}t \quad (3)$$

More generally, electrical energy can be measured by measuring voltage and current over a known amount of time. In practice, several decisions need to be made when developing an energy measurement approach. Firstly, the precise measurement of either current or voltage (they can be converted, as shown later), second, the integration of measured values and finally the reduction of the energy required for the actual sensing. The latter is necessary to reduce the influence that measurements have on the observed system. In the following, the observed device is referred to as DUT (device under test).

We now address each of these options in detail. In Section 2.2, we describe approaches to sense the influence that current, voltage and energy have on the environment. In Section 2.3, we discuss the means to amplify the sensed values, so that we can increase sensitivity for the purpose of reducing the measurement equipment's influence on the DUT. We will further talk about analog to digital conversion and the problems that arise from the discretization of time imposed by this conversion.

2.2. Sensing

A sensing unit is the basic element in a measurement setup. There are a variety of approaches to implement the sensing of energy consumption. In the following, the most prominent sensing approaches will be discussed.

2.2.1. Voltage drop or shunt measurement

The term "shunt" refers to an electrical resistor that is used to convert electrical current into a voltage drop for the purpose of measurement. In practice, voltage drop sensing is very common and easy to implement with analog to digital converters or even standard measurement-equipment such as a multimeter or oscilloscope.

According to Ohm's law, $V = R \cdot I$, a current I through a resistor R is caused by a voltage V proportional to R and I. Having a shunt resistor in series with a DUT results in an equal current flow through both components, while the supply voltage is split among them. This circuit of a *voltage divider* is schematically shown in **Figure 1**.

To avoid that the current being restricted mainly by the shunt and to avoid as a consequence the majority of the voltage drop occurring at the shunt resistor—both having negative impact on the DUT—the shunt resistance value must be chosen to be smaller by several magnitudes compared to the asserted resistance of the DUT.

As an example, we take a DUT that requires a constant voltage of about 4 V and a maximum current of 100 mA. This device has a theoretical maximum resistance of 40 Ω according to Ohm's law. With a badly selected shunt resistor of 4 Ω in series, the maximum current

http://dx.doi.org/10.5772/65989

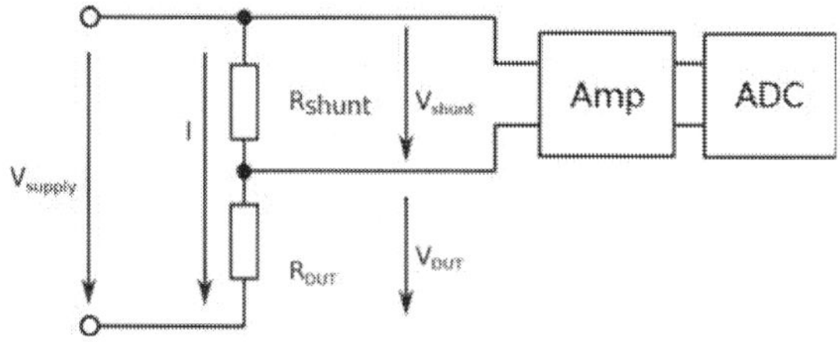

Figure 1. The concept of current measurement using a shunt in a voltage divider.

would be reduced to 91 mA. So, when the DUT is at maximum load, the voltage on the shunt resistor is 390 mV, whereas the supply voltage at the device drops to 3.7 V. This would not only distort the measurement results to a large degree but also could make the DUT malfunction.

On the other hand, using a shunt of 10 mΩ would just reduce the supply voltage to 3.999 V and the current to 99.98 mA. But, the maximum voltage to measure at the shunt would be below 1 mV.

In conclusion, the following assertions on shunts can be derived:

- The resistance value of the shunt is limited by the maximum acceptable voltage drop on the DUT's supply voltage and its maximum current consumption. This limits the measurement range.
- Usually, extremely small resistance values with low tolerances are preferable, which makes measurement shunts expensive.
- The small voltage drop at a shunt is prone to thermal noise.
- The small voltage drop at a shunt needs to be amplified to process the measurement signal.

To overcome these issues, there is a broad variety of high-quality (and high-cost) shunt resistors on the market to suit all measurement purposes. Additionally, there are technical approaches to enable precise or wide-range shunt measurements. To increase the measurement scale, some circuits use a shunt switching technology: whenever the voltage drop exceeds a threshold value, a lower Ohmic shunt is selected, thus enhancing the measurement range. This is not simple to implement because the transition from one shunt to the next must be exactly calibrated. Also, the threshold values for switching forward and backward must have a certain distance from each other (hysteresis) to prevent the system from oscillating between two shunts. Additionally, the current shunt selection must be communicated to successor circuits that process or use the measurements, since the evaluation of the measured voltage drop is dependent on the concrete value of the shunt.

Another approach often discussed is using exponential or logarithmic elements like diodes, instead of linear shunts. Usually these elements have technical issues such as strong temperature dependabilities and impractical tolerances that are hard to overcome.

2.2.2. *Charge accumulation*

Another method to measure energy is to use it to transfer an electrical charge into a capacitor. The charging capacitor will raise its voltage according to the following equation, with C being the capacity, measured in Farad, F.

$$E = \frac{1}{2} \cdot C \cdot V^2 \tag{4}$$

The energy stored in the capacitor can be derived from its voltage level. A good way to transport energy into a capacitor in proportion to the energy consumed by a DUT is the utilization of a *current mirror*. Current mirrors let a current pass through their input contacts while using an external power source to reproduce the same (or a proportional) current flow at their output side.

Assuming a constant voltage supply for the DUT, the *charge*, measured in Coulombs, C, which is equivalent to ampere seconds, As, transported to a connected capacitor is

$$Q(t) = Q(t_0) + \int_{t_0}^{t} I(t)\mathrm{d}t. \tag{5}$$

There are two suitable approaches to measure energy continuously: one is to sample and discharge the capacitor (or an alternating set of capacitors) with a constant frequency and the other approach is to wait until the capacitor voltage reaches a given limit to trigger a counter. **Figure 2** shows the corresponding circuit. In this approach, the frequency of the counter signal increases proportional to the energy consumption.

2.2.3. *Charge transfer*

Similar to accumulating charge, methods based on the transfer of charge into a capacitor can take the time into account that is required to transport energy into it. Capacitors are charged through a resistor. The time to increase the voltage of a capacitor is denoted by τ.

$$\tau = R \cdot C \tag{6}$$

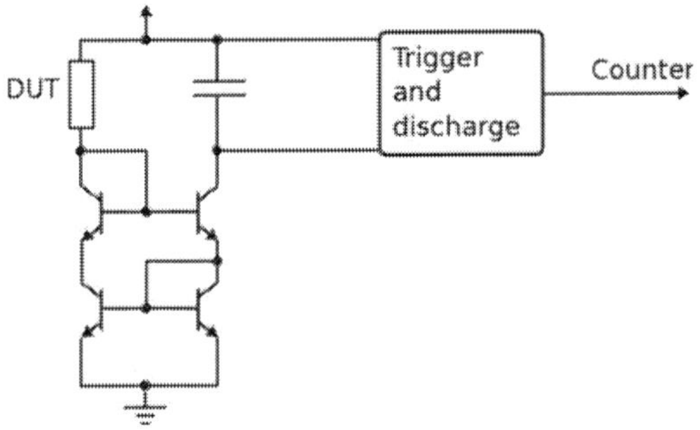

Figure 2. A Wilson current mirror used to charge a capacitor.

$$V(\tau) = V(t_0) + \left(V_{sup} - V(t_0)\right) \cdot \left(1 - e^{-1}\right) \approx V(t_0) + 0.632 \cdot \left(V_{sup} - V(t_0)\right) \tag{7}$$

Konstantakos et al. [5] have evaluated the usability of this measurement method for the in-situ power measurement of embedded systems. They implemented different circuits based on charge transfers using current mirrors and concluded that this method is accurate enough to serve as an in-situ measurement approach for embedded systems. Additionally, it can be implemented without dependencies to the clock frequency of the system.

One of the main advantages of using a capacitor to collect energy is the implicit integration of current over time. As capacitors are analog elements, there is no sampling involved, so that even very short energy peaks will be taken into account. A disadvantage is the additional analog circuitry required, which adds thermal noise and nonlinearities and thus is prone to reduce accuracy.

2.2.4. *Magnetic field*

A current flowing through a conductor creates a magnetic field around the conductor. This field can be sensed by devices like *Hall-effect sensors*. The Hall effect describes the occurrence of a voltage (Hall voltage) within a live conductor that is positioned perpendicular to an external magnetic field. That means, placing the live conductor to a DUT close-by and perpendicular to another live conductor (sensor) will induce a measurable Hall voltage into both conductors. Since both magnetic fields, that of the main conductor and that of the sensor, influence each other in the same way, the current through the sensor has to be smaller than the current through the examined conductor by several orders of magnitude.

The Hall effect has only a small impact on the voltage of the sensor, so a very sensitive amplifier has to be used to amplify its output signal. Hall-effect measurements are prone to errors, because many parameters influence the measurement. The conductor material, as well as its distance from the sensor and any insulating material between the two, all have a significant impact. Also, if there is air between the conductor and the sensor, air humidity and temperature might influence the measurement results.

The main advantage of Hall sensor measurements is the contact-free and nonintrusive way that a Hall sensor can be deployed. For these reasons, Hall sensors are mainly used to measure high currents in environments where invasive measurement is not desirable.

2.3. Signal processing

As shown in **Figure 1**, a typical measurement setup has to process the signal of the sensing unit to convert it into a human or machine readable form. This usually includes at least two stages: 1. A signal amplifier to adapt the output level of the sensing unit to satisfy the requirements of successor units. 2. Most often, the amplifier will be followed by an analog to digital conversion unit (ADC) to make the signal machine readable.

There is a broad range of measurement amplification circuits, and the quality and complexity of these circuits have an impact on the accuracy of the measurement and the effective measurement range. While simple units with low requirements may contain only a single bipolar transistor and few passive elements, more sophisticated setups that use one or more high-quality OpAmps allow for better temperature stability, better signal-to-noise ratio, amplification of both positive and negative signals and fewer parasitic effects like unwanted filtering of peaks, thus allowing measurements of signals of higher frequencies.

Among ADCs, the diversity is no less. Every ADC will convert a signal level (voltage) to a digital value by comparing the signal voltage to a reference voltage and calculating the ratio as digital value. Two fundamental parameters are the resolution, which is the number of bits per converted signal, and the sample frequency. Further parameters define the accuracy of the conversion, usually described by a set of error measures like offset error, gain error and nonlinearity in an ADC's data sheet.

An important design consideration is imposed by the Nyquist-Shannon sampling theorem, which limits the maximum signal frequency that can be sampled to half of the sampling frequency. Higher signal frequencies may cause errors in the conversion results. This adds the requirement to insert a low-pass filter into the signal path, which cancels out frequencies above the frequency limit. These filters are often integrated in ADC circuits and thus not considered any further in the design of measurement equipment. Due to the undefined quality of these filters, the input frequency range allowed in the product descriptions for commercial measurement equipment is often very limited.

Better filter systems integrate the signal accurately before sampling. Although low-pass filters and analog integrators are equivalent from a conceptual perspective, sophisticated integration units are more accurate and can limit the integration interval exactly to the sampling time. In the case of energy measurements, this would guarantee that the measurement result is always correct, even if the signal contains peaks of a width that is only a fraction the sampling interval time. Although such a peak would not be visible as such in the sample data, it would correctly increase the ADC's next output value. This concept was used in the MIMOSA measurement tool presented in Section 3.

3. Example direct measurement systems

This section presents two measurement systems designed to allow the energy consumption of embedded systems to be captured. The two systems, *MAGEEC wand* and *MIMOSA,* have slightly different design goals and use different measurement techniques. Both measurement systems offer fully working solutions; describing their construction here is expected to aid in the development of future measurement devices.

3.1. MAGEEC wand

The MAchine Guided Energy Efficient Compilation (MAGEEC) project[1] sought to find compiler optimizations that reduce energy. As part of this work, real hardware measurements were used, rather than model-based estimations. To that end, the MAGEEC wand, shown

[1]http://mageec.org/

http://dx.doi.org/10.5772/65989

in **Figure 3**, was created. Complementing the open source nature of the compiler work that was performed in the project, the wand's custom hardware and software is also open source. A number of MAGEEC wand kits were produced and both sold and given away at workshops. Using the published designs, anybody can commission the manufacture of their own boards.

3.1.1. *Features*

The MAGEEC wand is designed to be flexible in how it is used to suit various energy measurement needs. Its key features include:

- Three measurement channels that can be monitored simultaneously.
- Sample rates of up to 2 million samples per second.
- The measurement board attaches to a widely available, low-cost embedded system to capture and transmit collected data.
- It can use one of the available channels for self-monitoring of the capture device's energy consumption.
- Each channel can be easily adjusted to measure a wide range of power supplies and power ranges.
- Supplies with pre-installed shunt-resistors can also be monitored.
- The measurement firmware provides various capture methods, including streaming data, triggered capture and a live GUI.

Figure 3. MAGEEC wand attached to an STM32F4 discovery board.

3.1.2. Construction

The MAGEEC wand is a small add-on board designed to connect to an ST *discovery board,* which is a low-cost MCU evaluation board from ST that includes an ARM Cortex-M series processor. A block diagram of the relevant components is shown in **Figure 4**. The wand uses the common shunt-resistor based measurement method, as described in Section 2.2. It features a selection of shunt resistors per channel, probe points for connecting the DUT, inductors and an array of current sense amplifiers in a MAX4378FASD chip. The discovery board acts as the controller and data acquisition device. The on-chip ADCs are used to sample the voltage drop that was amplified on the wand. Sample data are delivered through USB to a connected PC.

Devices that use up to a 12 V power supply can be safely monitored with the wand. The power supply to the DUT may need to be modified to allow sensing, typically by splicing a cable. However, many devices, particularly evaluation boards of embedded systems, feature shunt resistors and probe points, removing the need for any hardware alterations. In the former case, an appropriate resistor value must be chosen, such that the voltage drop is sufficiently large to observe with minimal noise. Additionally, if the splicing is done by removing an inductor on the target board, an inductor on the wand can be used in its place. In the latter case, the inductor and resistors on the wand can be bypassed, although the shunt resistor value on the target device must be noted in order to correctly scale the measurements that are obtained.

The on-board resistors for each channel are 0.05 Ω, 0.5 Ω, 1 Ω and 5 Ω, with a header and space for a custom resistor per channel. Jumpers select the type of resistor to be used. A trigger pin can be assigned on the discovery board to allow sampling to be controlled by an external event, such as the toggling of a GPIO port on the DUT.

The design of the MAGEEC wand means that it does not provide power to a device. However, MIMOSA does, as described in Section 3.2, thus providing an alternative where this is preferred.

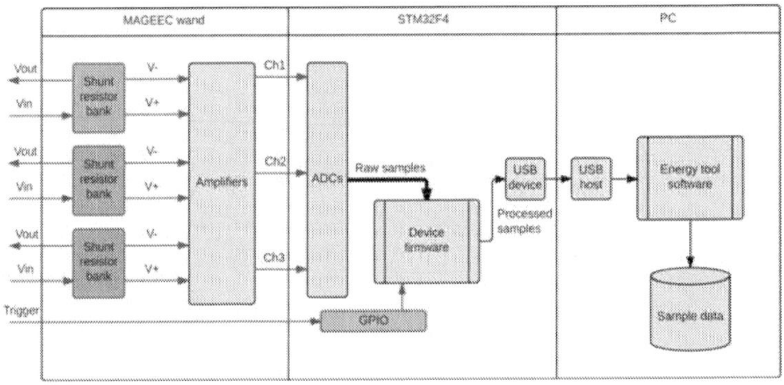

Figure 4. Block level depiction of MAGEEC wand and relevant STM32F4 components for taking energy measurements.

http://dx.doi.org/10.5772/65989

The device firmware and PC-side library allow data to be collected and presented in several ways. The *energy tool* program [6] is written in Python and can be run standalone or used as a library to integrate with other tools. It supports three main modes of acquisition:

1. In triggered mode, the device firmware samples power during the triggering period of the assigned GPIO. At the end of the triggering period, the duration, average and peak power, as well as total energy, are provided to the host PC.

2. In continuous mode, data are continuously sampled. The firmware aggregates samples to provide data to the USB host at a reasonable data rate.

3. In interactive mode, continuously sampled data are provided in a real-time graph. Parameters such as channel selection and resistor values can be interactively configured. This provides easy experimentation and observation prior to programmatically configuring these parameters for automated data collection.

An additional bundled tool, *platformrun*, combines the energy tool with the ability to run programs on a DUT, coupling the compilation and execution of a program of interest with the data collection process. This allows automated collection of program energy consumption under changing parameters, such as the compiler parameters explored by the MAGEEC project.

3.1.3. Uses

The MAGEEC wand has been used in various contexts to date. In 2014, a FOSDEM workshop was held where attendees could set up a wand to measure the energy of a variety of embedded devices.[2]

In research, the wands have been used to collect energy data for processor and communication modelling [7], data-dependent energy modelling [8, 9] and exploration of compiler optimizations for energy efficiency. Finally, the MAGEEC wand was of course pivotal in the research output of the MAGEEC energy efficient compiler optimization research.

3.2. MIMOSA

Buschhoff et al. [10] proposed MIMOSA[3], a measurement device for creating high-accuracy energy models of embedded system components. MIMOSA combines different measurement approaches. It acts as the power source for the DUT, and by that it measures the energy that it delivers to the DUT. This is achieved by implementing a constant voltage source using a feedback loop on an operational amplifier (OpAmp) to create the output voltage from a high-impedance reference voltage. Such a circuit is sometimes referred to as *voltage follower*.

Compared to a typical voltage follower, which feeds back the OpAmp's output to one of its inputs directly, MIMOSA breaks the feedback loop with a transistor, as shown in **Figure 5**. Since the OpAmp strives to cancel out the voltage difference on its input pins, the transistor

[2]MAGEEC FOSDEM workshop: http://mageec.org/wiki/Workshop

[3]"Messgerät zur integrativen Messung ohne Spannungsabfall," German for "Measurement device for integrative measurements without voltage drop."

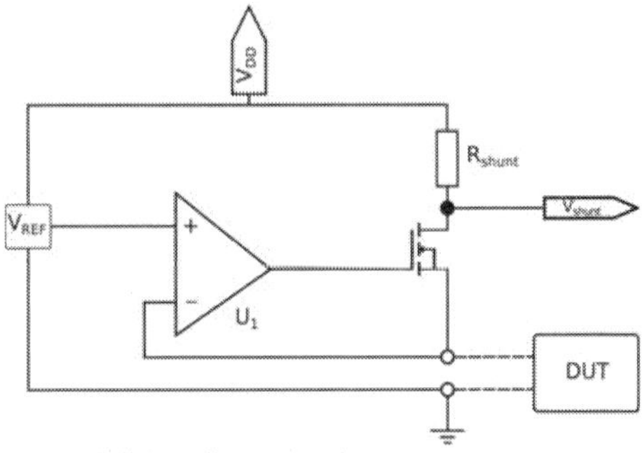

Figure 5. MIMOSA voltage regulator circuit.

will be forced to create the reference voltage at the DUT side by the OpAmp, whereas the current supply for the DUT actually has to come over a shunt from MIMOSA's own voltage supply. For that to work, the MIMOSA supply voltage must be greater than that of the DUT. Contrary to normal shunt measurement, the shunt used here can be high ohmic because the voltage drop of the shunt is compensated by the regulatory circuit.

Using a high-ohmic shunt here has some advantages over the usual "shunt plus amplifier" approach. As an example, a current of 1 mA would create a voltage drop of 1 V at a 1 kΩ shunt. That means, no further amplification is necessary, and a standard ADC connected to the shunt can sense currents down to the micro ampere region. Due to the fewer components required, thermal noise is less of an issue, while precise high-ohmic resistors are far less expensive than precise measurement shunts in the milliohm-range. **Figure 6a** shows a reference measurement in a DC situation. Ten high precision ohmic resistors were used as load. For each resistor, 300,000 measurements were taken. The figure shows the measured range as a vertical bar for each resistor. **Figure 6b** shows the distribution of the measurement results (normalized average to the linear regression straight).

In the next stage, MIMOSA integrates the voltage measured at the shunt by using a set of three analog integrators (each consisting of an OpAmp and a capacitor). This can be seen in the overview of the system in **Figure 7**. The integrators function in a rotational manner: while one integrator is connected to the shunt, another one's output value is sampled by the ADC and the third integrator is reset. The sampled value is then sent to a PC using a USB connection.

MIMOSA additionally features a digital input that can be used to tag the collected samples, bookmarking important events. The DUT can use this connection to signal important events like the start and the end of a program sequence to analyze. The state of the digital input is encoded within the data stream.

On the connected PC, data can be evaluated with a graphical user interface. The user can display and record measurements, select and cut interesting sections and export data for further evaluation (see **Figure 8**).

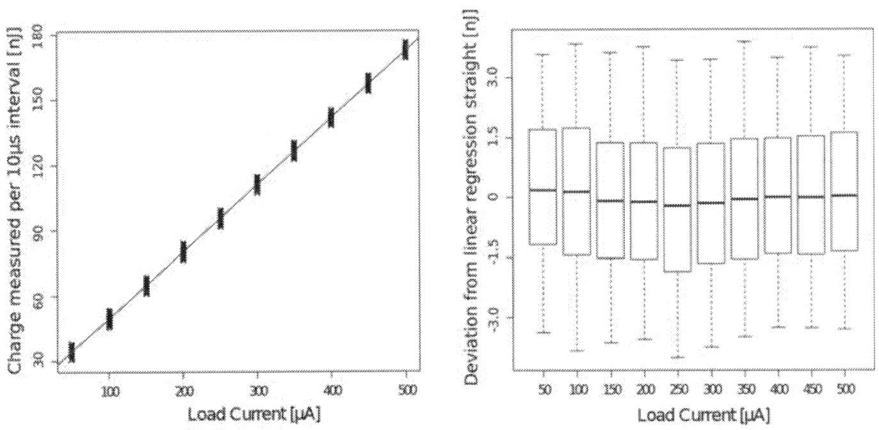

(a) DC linearity (b) Noise

Source: [10], with permission of Springer

Figure 6. DC measurement with 10 precision resistors [10]. (a) DC linearity and (b) noise.

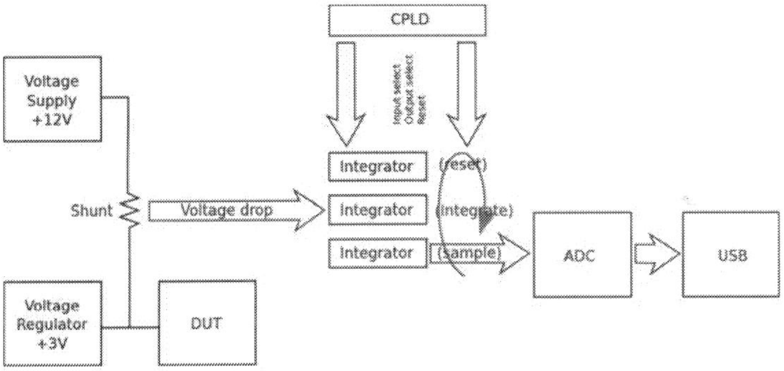

Source: [10], with permission of Springer

Figure 7. Overview of the MIMOSA architecture [10].

Figure 9a shows measurements of rectangular signals that are shorter than the actual sampling period of 10 µs at a sampling frequency of 100 kHz. Throughout the measurement range, MIMOSA shows results close to linear. This is depicted more precisely by the box plot in **Figure 9b**, where the deviation from the regression line and the distribution of measurement values is shown.

In conclusion, MIMOSA aims at the precise measurement of deeply embedded systems, as it allows for the sensing of current in the lower μA region, while still having good time precision through its sample rate of 100 kHz. Due to the integration circuits, MIMOSA guarantees a good energy measurement accuracy; even small peaks far below the sample period will be accounted for. The sample rate can be raised by using higher value sampling

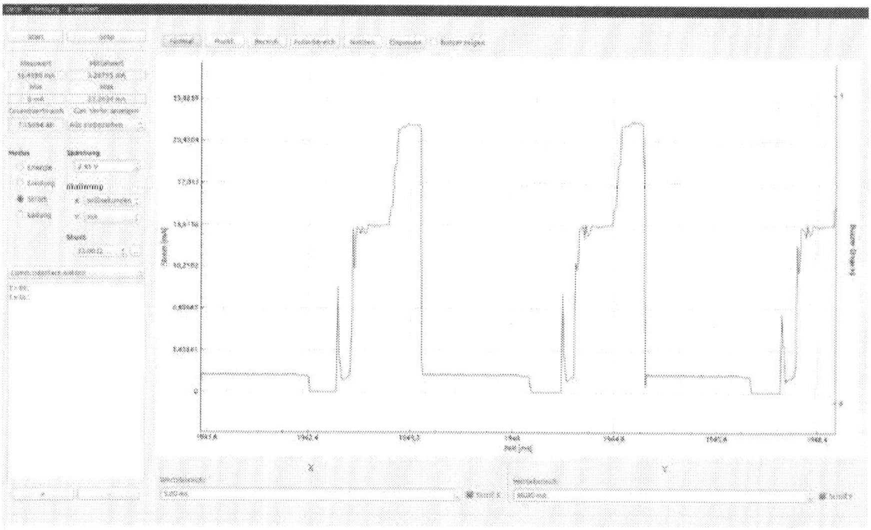

Figure 8. The MIMOSA GUI application.

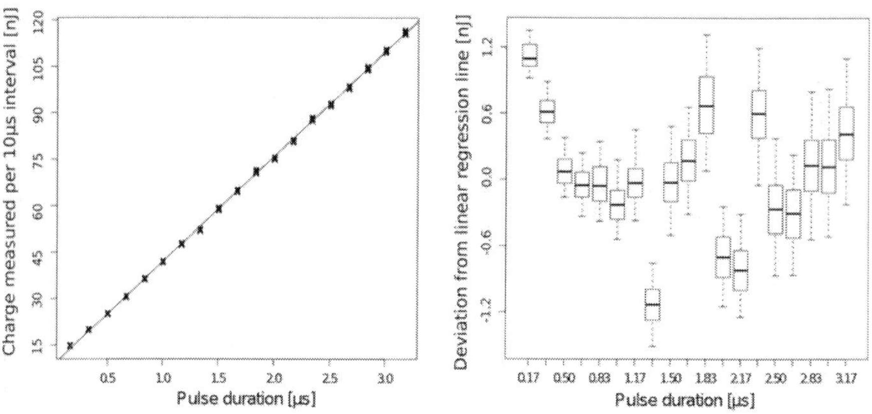

(a) Linearity

(b) Noise

Source: [10], with permission of Springer

Figure 9. Integration of peaks smaller than the sample period (10 μs) [10]. (a) Linearity and (b) noise.

devices like oscilloscopes right behind the sensing unit. In comparison to existing commercial devices for energy measurement in embedded systems, MIMOSA is able to represent the signal form more precisely without loss of overall accuracy [10]. On the downside, MIMOSA requires a more sophisticated setup as it is necessary to replace the constant voltage source of the DUT.

http://dx.doi.org/10.5772/65989

4. Basics of regression-based techniques

Modifying and instrumenting hardware to measure its energy consumption may be difficult or undesirable in some circumstances. A certain level of skill is required, and it may not be practical to provide the measurement apparatus to all instances of a device. An alternative is to construct a model that provides measurements using alternative data sources as a proxy. This still yields run-time energy samples, but they are sourced indirectly.

This section explains how regression-based techniques can be used to establish the parameters necessary to extract energy consumption from other metrics. The subsequent section then demonstrates an application of this.

4.1. What is regression analysis?

Regression is a method to investigate the functional relationship among variables. The relationship is expressed in the form of an equation or a model connecting the *response* (or *dependent*) variable and one or more *predictor* (or *explanatory*) variables.

The response variable is denoted using Y and the set of predictor variables by $X_1, X_2, \ldots, X_p$, where p denotes the number of predictor variables. The relationship between Y and the set of predictor variables X can be expressed by a general regression model,

$$Y = f\left(X_1,\ X_2, \ldots,\ X_p\right) + \epsilon, \tag{8}$$

where ε is assumed to be a random error representing the discrepancy in the approximation. The function f describes the relationship between Y and the predictor variables in X. An example of a linear regression model is.

$$Y = \beta_0 + \beta_1 X_1 + \beta_2 X_2 + \ldots + \beta_p X_p + \epsilon, \tag{9}$$

where β variables are called regression parameters or regression coefficients, which are unknown constants to be determined (estimated) from the data.

We can decompose the problem of practical regression analysis into the following six steps:

1. Statement of the problem.
2. Selection of potentially relevant variables.
3. Data collection.
4. Model specification.
5. Model fitting.
6. Model validation.

The first three steps require a good understanding of the problem so that a good selection of the predictor variables can be made. These should be variables with a strong relation with the response variable. For example, if trying to model power consumption in a CPU, the operat-

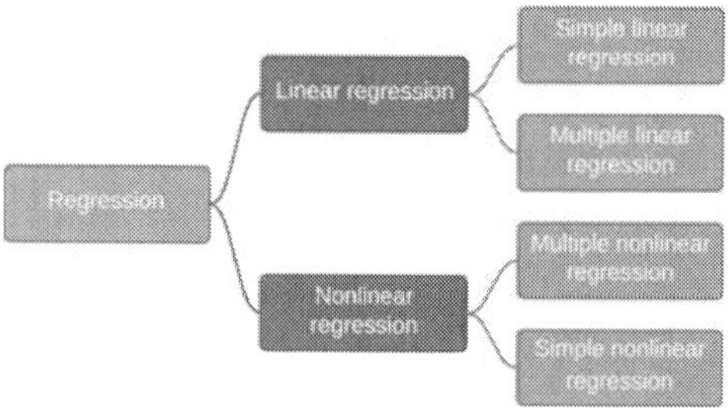

Figure 10. Various categorizations of regression approaches.

ing frequency and voltage will be good predictors to start with together with the number of cycles spent in different processor states, such as idle, active, etc. Others could include the type of instructions (floating point, arithmetic, load/store, etc.), or data on micro architectural events such as the cache miss rate. A good understanding of the problem is important because the effects of some predictors could be unexpected. For example, a high cache miss rate is not desirable, but it will result in an instantaneous power reduction since the core will spend more cycles doing nothing while waiting for data. The data collection step should exercise the predictor variables as thoroughly as possible so that enough data is collected to link the predictors and response variables. Once this data has been collected, the following steps can be performed.

4.2. Model specification

To specify the model, we need to decide what type of regression we are going to use. Regression is divided into two basic types, linear regression and nonlinear regression (see **Figure 10**). Linear regression can be performed with simple linear regression or multiple linear regression and the same applies to the nonlinear case. To solve linear regression, the least squares method is most often used. However, to solve nonlinear regression, variable transformation must be applied on the nonlinear model at first [11, pp. 1405–1411] to obtain a linearized model and then the linear methods can be used with the new linear model.

For example, each of the four following models is linear:

$$Y = \beta_0 + \beta_1 X + \epsilon \tag{10a}$$

$$Y = \beta_0 + \beta_1 X_1 + \beta_2 X_2 + \epsilon \tag{10b}$$

$$Y = \beta_0 + \beta_1 \log X + \epsilon \tag{10c}$$

$$Y = \beta_0 + \beta_1 \sqrt{X} + \epsilon \tag{10d}$$

because the model parameters β are related linearly to the response variable Y. On the other hand,

$$Y = \beta_0 + e^{\beta_1 X} + \epsilon \tag{11}$$

is not a linear model because the coefficient β_1 does not enter the model linearly and the relationship between Y and X is not linear either. To satisfy the assumption of the standard linear regression model, it is sometimes possible to apply an appropriate transformation of the variables to the equation so that the relationship between the transformed variables and the new response variable becomes linear. Instead of working with the original variables, working with the new transformed variables linearizes the model and thus simplifies the approach. Taking Eq. (11) as an example, after applying logarithm on both sides, the equation turns into:

$$\ln Y = \ln \beta_0 + \beta_1 X + \epsilon \tag{12}$$

Now the response variable, ln Y, has a linear relationship with the predictor variable, X, and coefficient, β_1.

As an example, the model equation in a power model could be:

$$P = aV^2 f + bV^3 + cV^5 \tag{13}$$

In Eq. (13), the first term represents dynamic power; the second term is subthreshold leakage and the third term is gate leakage. The gate leakage tends to be very small when compared with the prior two terms and tends to be disregarded. The equation is linear with multiple nonlinear variables, but it does not violate the assumption of a standard linear regression model because coefficients have already entered the equation linearly. Hence, the simplest way is to regard the terms V^2f and V^3 as predictor variables X_1 and X_2, respectively, and then the original model in Eq. (13) becomes a multiple linear regression model, formally described by Eq. (14).

$$P = aX_1 + bX_2 \tag{14}$$

4.3. Model fitting

Model fitting estimates the regression parameters or, in other words, fits the model to the collected data using the chosen estimation method. The estimates of coefficients β_0, β_1, ..., β_p are denoted by $\hat{\beta}_0, \hat{\beta}_1, \ldots, \hat{\beta}_p$. In the case of linear regression, the estimated regression equation then becomes

$$\hat{Y} = \hat{\beta}_0 + \hat{\beta}_1 X_1 + \hat{\beta}_2 X_2 + \ldots + \hat{\beta}_p X_p. \tag{15}$$

The value $\hat{Y}$ is called the fitted value [12] computed using estimated parameters and the values of predictor variables in the observation. Using Eq. (15), we can compute n fitted values of Y for n observations in the data set. Hence, the fitted value $\hat{Y}_i$ in the ith observation is

$$\hat{Y} = \hat{\beta}_0 + \hat{\beta}_1 x_{i1} + \hat{\beta}_2 x_{i2} + \ldots + \hat{\beta}_p x_{ip}, \quad i = 1,2,\ldots,n. \tag{16}$$

where $x_{i1}, x_{i2}, \ldots, x_{ip}$ are the values of the p predictor variables for the ith observation. These fitted values are the quantities needed for computing *correlation* which can evaluate the goodness of the model. In addition, we can use them to compare with the real value of the response variable y_i and compute the errors (or residuals) in order to evaluate the goodness of the fitting in a different way. This provides the average error which should be within tolerable bounds for the regression to be successful.

To solve simple linear regression, the least squares method is commonly used. It is possible to compute the regression coefficients with scripts for Matlab or Octave or even with a spreadsheet application. Based on the available data, we aim to estimate the value of the regression parameters and find a straight line (or surface for multiple linear regression) that gives the best fit. A best fit means that this fitting could give the smallest sum of squares of errors (residuals). The smallest sum of square of errors is obtained by minimizing the sum of squares of the *vertical distances* from each point to the line. These vertical distances represent the errors between the estimated response variable and the real response variable. Rearranging the equation of a standard linear model as given in Eq. (9), the errors are represented as

$$\epsilon_i = y_i - \beta_0 - \beta_1 x_i, \quad i = 1, 2,\ldots,n \tag{17}$$

The sum of squares of errors (SSEs) can then be written as

$$S(\beta_0, \beta_1) = \sum_{i=1}^{n} \epsilon_i^2 = \sum_{i=1}^{n} (y_i - \beta_0 - \beta_1 x_i)^2 \tag{18}$$

The values of the two coefficients that minimize the SSE value are given by using the least squares method [13]

$$\hat{\beta}_1 = \frac{\sum (y_i - \bar{y})(x_i - \bar{x})}{\sum (x_i - \bar{x})^2} \tag{19}$$

and

$$\hat{\beta}_0 = \bar{y} - \hat{\beta}_1 \bar{x} \tag{20}$$

where

$$\bar{x} = \frac{\sum_{i=1}^{n} x_i}{n} \quad \text{and} \quad \bar{y} = \frac{\sum_{i=1}^{n} y_i}{n} \tag{21}$$

http://dx.doi.org/10.5772/65989

are the mean of the response variable and predictor variable, respectively. The estimates $\hat{\beta}_0$ and $\hat{\beta}_1$ are called least squares estimates of β_0 and β_1, respectively. They represent the intercept and slope of the line that gives the minimum sum of squares of the vertical distance from each observation point to the line. This line is called the least squares regression line because it is the solution to the least squares method. The least squares regression line is formalized as

$$\hat{Y} = \hat{\beta}_0 + \hat{\beta}_1 X \tag{22}$$

To compute each fitted value in each observation:

$$\hat{y}_i = \hat{\beta}_0 + \hat{\beta}_1 x_i, \quad i = 1, 2, \ldots, n \tag{23}$$

The vertical distance corresponding to the *i*th observation is

$$\epsilon_i = y_i = \hat{y}_i, \quad i = 1, 2, \ldots, n \tag{24}$$

This kind of vertical distances is called *ordinary least squares residual*. These residuals have satisfied the property that their sum is zero, which means that the sum of distances of the points above the line is equal to the sum of the distances below the line.

4.4. Model evaluation

After fitting a regression model relating the response variable with the predictor variables, it is important to determine the quality of the fit. Covariance and correlation measure the direction and strength of the linear relationship between the two variables (response variable and predictor variable). The quantity correlation is key to evaluate the goodness of the fit. An additional useful measure of the quality of the fit is the R-square value. To obtain the R-square value, there are three quantities that need to be computed:

$$\text{SST} = \sum (y_i - \bar{y})^2 \tag{25}$$

$$\text{SSR} = \sum (\hat{y}_i - \bar{y})^2 \tag{26}$$

$$\text{SSE} = \sum (y_i - \hat{y}_i)^2 \tag{27}$$

where SST denotes the sum of squares of the deviations in Y from its mean, SSR represents the sum of the squares due to the regression and SSE stands for the sum of squares of residuals (errors). To understand the significance of these measurements, we can write the following simple relation between observed values and estimated values:

$$\begin{array}{cccc} y_i = & \hat{y}_i & + & (y_i - \hat{y}_i) \\ \text{Observed} = & \text{Fit} & + & \text{Deviation from fit,} \end{array} \tag{28}$$

then subtracting $\bar{y}$ from both sides of Eq. (28), it becomes

$$\begin{array}{rccc} y_i - \bar{y} = & (\hat{y}_i - \bar{y}) & + & (y_i - \hat{y}_i) \\ \text{Deviation from mean} = & \text{Deviation due to fit} & + & \text{Residuals} \end{array} \quad (29)$$

It is obvious that the total sum of squared deviations in Y, SST, is decomposed into two parts: one is the SSR, which measures the quality of X as a predictor of Y, and the other one is SSE that measures the error in the estimation. The quantity R-square value is the ratio of SSR to SST which tests how accurate the fit is; it removes the contribution of residuals in the SST. Hence, the R-square value, R^2, is written as

$$R^2 = \frac{\text{SSR}}{\text{SST}} = 1 - \frac{\text{SSE}}{\text{SST}}. \quad (30)$$

If the residuals corresponding to the SSE are 0, then the estimation is perfect and the R-square value is 1. Therefore, the closer it is to 1, the stronger the fit.

5. Applying linear regression to estimate power requirements of an ARM processor

The evaluation board uses an ARM Cortex A9 processor equipped with a PMU (performance monitoring unit) that can monitor six performance counters out of a maximum of 58 available events. The event profiling is based in the Linux utility *perf* configured to only monitor events corresponding to the application being executed. We focus on activities related to cache misses, instruction execution and CPU states that have been shown to have a strong influence on power. However, the hardware limitation means that the model is limited to a maximum of six coefficients. The benchmark selected is Mibench [14] and the whole benchmark is divided into two groups so that one group is used for model training and the other group is used for model verification.

The instructions that can be monitored by the PMU are integer instructions, load/store instructions and floating-point instructions. Our experiments show that the integer clock enable state and the data engine clock enable estate have a strong correlation with the integer instructions and offer more accuracy in the model than directly counting the number of integer instructions. On

Predictor variables	Coefficient values
Instruction cache miss	-5.4683×10^{-8}
Data cache miss	-1.4589×10^{-8}
Load/store instructions	4.7787×10^{-11}
Floating-point unit instructions	2.5745×10^{-9}
Integer clock enabled	3.6552×10^{-10}
Data engine clock enabled	3.7001×10^{-11}

Table 1. Coefficients for the final model.

http://dx.doi.org/10.5772/65989

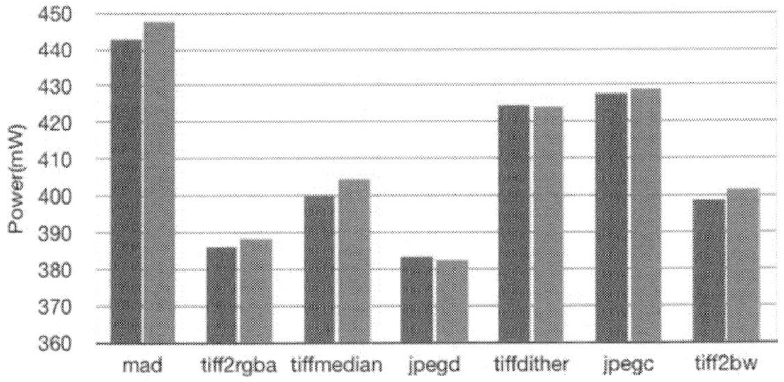

Figure 11. Estimated (blue) and measured (green) average power for the model with Mibench applications.

the other hand, the floating-point instructions and the load/store instructions show a high correlation between the estimated power and the measured power and for this reason are selected.

Both the instruction cache misses and the data cache misses significantly influence power usage since they result in stalls in the pipeline while the data is fetched from main memory. For this reason, the model coefficients shown in **Table 1** have negative values associated with the cache misses. This means that these cache misses result in a reduction in processor power due to additional stalling. Notice that overall, this is not a positive effect since the cache misses will increase power in the memory subsystem and also increase execution time resulting in an overall increase in energy usage.

The values shown in **Table 1** correspond to the coefficients $\hat{\beta}_0, \hat{\beta}_1, \ldots, \hat{\beta}_6$ according to Eq. (16) in Section 4.3, while the six events shown in **Table 1** are the predictors. The final constant value $\hat{\beta}_0$ represents idle power or power that is not due to execution of the application. We have measured idle power at a value of 356 mW, and this includes leakage, clock network power and some overhead power due to the Linux OS. **Figure 11** evaluates the goodness of the model with Mibench applications showing that a relative simpler linear model can achieve useful accuracy within 5–10% of measured values.

6. Summary

To reduce energy consumption, it is necessary to understand how much energy a device consumes. Measurement methods are therefore essential not just to help designers and developers understand the behavior of their devices, but also to help them gauge the success of any energy-saving efforts that they make. This chapter has presented several measurement approaches, each with a unique set of properties and potential use cases.

The first general approach, direct measurement, was presented in Section 2. This involves hardware that can detect energy consumption, for which there are several methods, including current sensing, charge accumulation or transfer measurement and magnetic field sensing. Each measurement circuit has its own level of complexity, precision, range and level of invasiveness with respect to the target hardware. Two example measurement systems are presented in Section 3 and their respective capabilities discussed.

Beyond direct methods of measurement, there is model-assisted measurement. In Section 4, regression analysis methods are presented, which can be used to estimate a device's energy consumption through other properties that can be noninvasively observed, such as performance counters or other events that take place during program execution. Successful measurement of this kind requires that the parameters of the model are carefully selected and understood. Following a demonstration of the construction of a model, an example of a linear regression model, applied to an ARM Cortex A9 processor, is given in Section 5.

Choosing the best approach is dependent upon the device or devices to be measured, as well as the use case and intended outcome. For example, deeply embedded devices may have fewer sources of data to inform a linear regression model, thus direct measurement may be preferred. Or, the deeply embedded processor may be sufficiently simple that an equally simple linear regression model is acceptable. In a more complex system, there may be a desire to know exactly where energy is being consumed, for example, in RAM, buses or a particular type of computation. Without sufficient prior knowledge, this may be difficult to understand without multiple direct measurement points.

Regardless of the measurement method, care must be taken to control, or account for, external factors. An example of this is temperature. The temperature of a device affects its leakage current and therefore its energy consumption. Its temperature will be governed by how active the device is, as well as the ambient temperature, and the ability of the system to remove heat from the device, be it passively or actively. The efficiency of power supplies is also governed in part by temperature as well as load level. Thus, over time, and in different environmental conditions, energy measurements may change for an otherwise unchanged use case.

The permanence, transferability and side-effects of the measurement setup must also be considered. A noninvasive approach can be used on any instance of a device, maximizing transferability, allowing run-time monitoring of a device in a broad set of scenarios. However, this may come with a loss of accuracy, and if the data collection is introspective—collected and processed on the device under test—then the overhead of this effort must also be accounted for. Higher precision typically requires higher effort, as well as additional supporting hardware, removing processing overheads from the device under test, but placing an overhead on the developer in terms of additional equipment, tooling, data collection and analysis. A high sampling rate may seem desirable, but the increase in data collected may be excessive. Devices that are permanently tooled for energy measurement may not be desirable, as in some scenarios, the energy consumption of monitoring will itself be a problem.

Key questions in selecting a measurement method

To summarize, the following questions should be posed in order to guide the selection of an appropriate energy measurement setup.

http://dx.doi.org/10.5772/65989

How many devices need measuring? A one-off tooling versus one that needs replicating many times will influence the preferred method.

What level of detail is required? Establish whether multiple measurement points are necessary, or if a single total energy is sufficient, and what level of precision and accuracy are needed.

What is the overhead? Can the device under test tolerate additional processing burden, or must the data be collected and processed externally? The effort required by the software engineer must also be considered.

Where will the data be used? Run-time decision making requires always-available data, whereas data used to improve a system in development is no longer needed once the product is shipped.

Are there uncontrollable factors? Environmental conditions such as temperature can affect energy, and if they are not controlled, useful data cannot be obtained. Similarly, if the typical operating environment is volatile, then it is more difficult to make concrete assumptions based on data collected in a more controlled (for example, lab or workbench) environment.

Acknowledgements

This chapter is partly based on work performed in the EU 7th Framework Programme project ENTRA: Whole-Systems ENergy TRAnsparency (318337) and the MAGEEC project, supported by the Technology Strategy Board of the UK government under its Energy Efficient Computing Initiative. This work was also partly supported by the German Research Council (DFG) within the Collaborative Research Center SFB 876, project A4.

Author details

Steve Kerrison[1], Markus Buschhoff[2], José Nunez-Yanez[1] and Kerstin Eder[1,*]

*Corresponding author E-mail: kerstin.eder@bristol.ac.uk

1 University of Bristol, UK

2 TU-Dortmund University, Germany

References

[1] C. Edwards. Lack of Software Support marks the Low Power Scorecard at DAC. Electronics Weekly, 2011, No. 2472, p 6.

[2] R. Rodrigues, A. Annamalai, I. Koren, and S. Kundu. A study on the use of performance counters to estimate power in microprocessors. IEEE Transactions on Circuits and Systems II: Express Briefs, 60(12):882–886, 2013.

[3] Intel Corporation. Intel Xeon Phi Coprocessor (codename: Knights Corner) Performance Monitoring Units. Intel Corporation, Report number: 327357-001, 2012.

[4] ARM Holdings. Cortex-A9 Technical Reference Manual, revision r2p0. ARM Holdings. Report number: DDI0388E, 2009.

[5] V. Konstantakos, K. Kosmatopoulos, S. Nikolaidis, and T. Laopoulos. Measurement of power consumption in digital systems. IEEE Transactions on Instrumentation and Measurement, 55(5):1662–1670, 2006.

[6] J. Pallister. STM32F4 Energy Monitor. Available from: https://github.com/jpallister/stm32f4-energy-monitor [Accessed Nov. 2016].

[7] S. Kerrison and K. Eder. Modeling and visualizing networked multi-core embedded software energy consumption. Computing Research Repository (CoRR). Report number: abs/1509.02830, 2015.

[8] J. Morse, S. Kerrison, and K. Eder. On the infeasibility of analysing worst-case dynamic energy. Computing Research Repository (CoRR). Report number: abs/1603.02580, 2016.

[9] J. Pallister, S. Kerrison, J. Morse and K. Eder. Data dependent energy modelling: a worstcase perspective. Computing Research Repository (CoRR). Report number: abs/1505.03374, 2015.

[10] M. Buschhoff, C. Günter and O. Spinczyk. MIMOSA, a highly sensitive and accurate power measurement technique for low-power systems. In Real-World Wireless Sensor Networks, Lecture Notes in Computer Science. Springer, Berlin Heidelberg, 2013.

[11] G.K. Smyth. Nonlinear regression. In: Gudmund Høst (ed.), Statistical and Numerical Computing, Encyclopedia of Environmetrics, 4, John Wiley & Sons Ltd, 2006.

[12] S. Chatterjee and A.S. Hadi. Regression analysis by example. John Wiley & Sons, 2015.

[13] E.W. Weisstein. Least squares fitting. Available from: http://mathworld.wolfram.com/LeastSquaresFitting.html [Accessed Nov. 2016]

[14] M.R. Guthaus, J.S. Ringenberg, D. Ernst, T.M. Austin, T. Mudge, and R.B. Brown. Mibench. A free, commercially representative embedded benchmark suite. In Proceedings of the Workload Characterization, 2001. WWC-4. 2001 IEEE International Workshop, WWC '01, pp. 3–14, Washington, DC, USA, 2001. IEEE Computer Society.

Chapter 4

An Energy-Efficient Design Paradigm for a Memory Cell Based on Novel Nanoelectromechanical Switches

Azam Seyedi, Vasileios Karakostas,
Stefan Cosemans, Adrian Cristal,
Mario Nemirovsky and Osman Unsal

Additional information is available at the end of the chapter

http://dx.doi.org/10.5772/105947

Abstract

In this chapter, we explain NEMsCAM cell, a new content-addressable memory (CAM) cell, which is designed based on both CMOS technologies and nanoelectromechanical (NEM) switches. The memory part of NEMsCAM is designed with two complementary nonvolatile NEM switches and located on top of the CMOS-based comparison component. As a use case, we evaluate first-level instruction and data translation lookaside buffers (TLBs) with 16 nm CMOS technology at 2 GHz. The simulation results demonstrate that the NEMsCAM TLB reduces the energy consumption per search operation (by 27%), standby mode (by 53.9%), write operation (by 41.9%), and the area (by 40.5%) compared to a CMOS-only TLB with minimal performance overhead.

Keywords: nanoelectromechanical (NEM) switch, content-addressable memory (CAM) cell, translation lookaside buffer (TLB)

1. Introduction

Computing technology has witnessed an inimitable progress in the last decades, which is the result of CMOS technology scaling commensurate with Moore's law [1]. Transistor feature sizes have shrunk to half at each generation, and consequently, the number of transistors per chip has doubled every 2 years. However, CMOS scaling faces serious problems that occur due to the exponential increase of the leakage of current during technology scaling [2]. The subthreshold leakage is mainly affected by the subthreshold swing (S) of a device, which is defined as the amount of gate voltage reduction to reduce the subthreshold current by one decade (S = dVgs/dlogId) [3]. For bulk CMOS, the subthreshold swing has a substantially lower limit of 60 mV/decade, which leads to a large increase in the power density [4]. This limitation prevents manufacturers from fabricating smaller devices and forces them to look for alternative solutions targeting higher performance and efficiency. In order to maintain the scaling ability, a significant amount of research is ongoing to explore various non-CMOS technologies (emerging technologies) as a replacement for volatile and nonvolatile memories. This also motivates us, in this chapter, to exploit one of the most promising emerging technologies (nanoelectromechanical switches) to improve the processor performance.

Nanoelectromechanical (NEM) switches have been suggested as a promising candidate for replacing the CMOS technology [5]. NEM switches provide some unique characteristics that are not available in conventional MOS, such as near-zero leakage current and infinite subthreshold slope (<0.1 mV/decade [6]). Such characteristics make them ideal for designing highly energy-efficient structures. However, NEMs have relatively long mechanical switching delay [5] compared to the intrinsic delay of CMOS devices, and to this date, they suffer from low endurance (10^{11} write cycles) [7]. Also, NEMs do not offer high turn-on current like CMOS transistors.

In spite of the large mechanical delay and limited number of reliable cycles, NEMs have been useful for a wide range of applications such as FPGAs (used as programmable routing switches) [8], adders [9], flip-flops [10], memories [4, 11], DACs [9], and ADCs [9], where the long switching time and limited number of hits are not important issues. Most of the mentioned circuits use the benefits of combining NEMs and CMOS technology in order to highlight the advantage points of each technology and alleviate the disadvantages to achieve low-power and high-performance operation for some critical components. Motivated by these observations, in this chapter, we describe a new content-addressable memory (CAM) cell design, NEMsCAM [12], based on both NEMs and CMOS technologies, to employ in processor structures where writes are relatively infrequent, for example, the translation lookaside buffers (TLB).

Content-addressable memories (CAM) have been widely adapted for applications that depend on fully associative and high-speed search operations, such as translation lookaside buffers (TLBs), network routers, and data compression [13]. Since the search operation requires fully parallel and fast comparisons, CAMs introduce high energy consumption and area constraints. Previous works have explored the design of CAMs with emerging memory technologies to mitigate these issues [14–18]. However, those CAMs suffer mainly from increased search latency due to the employed technology that prevents them from building performance critical structures such as TLBs.

The memory component of NEMsCAM cell is designed with two complementary nonvolatile NEM switches while the comparison circuits are designed with CMOS transistors to allow fast search operation. The novel structure of NEMsCAM along with unique characteristics of NEMs considerably reduces the layout area as well as the energy consumption. Also, in this design, both Out and OutB are available simultaneously, which is essential to design a CAM cell.

As a use case, we leverage the NEMsCAM cell to build fully associative TLBs. The TLB has been pointed out as a critical component of energy and performance in modern processors [19]. Translation lookaside buffer (TLB) is a cache that is employed to accelerate virtual-to-physical address translation [20]. The processor searches the TLB on every memory operation using the virtual page number. Hence, the TLB is a crucial component for the performance and power consumption of the computers [20]. We evaluate first-level data and instruction TLBs with 16 nm technology at 2 GHz frequency. Our analysis demonstrates that the proposed TLB reduces the energy per search and write operation and standby mode by 27%, 41.9% and 53.7%, respectively, and the area by 40.5% compared to a CMOS-only TLB. Also, both designs execute the search operation in one clock cycle. Furthermore, it is shown that the NEMs' increased write latency introduces minimal performance overhead (0.27% on average). The main contributions of this chapter are as follows: (1) Description of NEMsCAM cell design based on complementary nonvolatile NEM switches and CMOS transistors. (2) Explain the design of highly efficient first-level TLBs for data and instruction accesses based on the NEMsCAM cell. (3) Evaluate the proposed designs at both circuit and system level and compare to CMOS-only TLBs.

Section 2 provides background information whereas Section 3 and Section 4 describe the design of the NEMsCAM cell and the NEMsCAM TLB, respectively. In Section 5, we present our evaluation methodology and the obtained results. Finally, in Section Section 6, we summarize the chapter.

2. Background

This section provides background information related to this work. First, we briefly review recent available emerging technologies. Then, we describe NEM switches in detail and the prior art in their use of memory. Finally, we describe CMOS-only CAM cells and fully associative TLB structures.

2.1. Emerging technologies

In order to continue the trend of Moore's Law, many emerging technologies have been employed such as phase-change memory (PCRAM) [21], magnetoresistive RAM (MRAM) [22], spin-torque transfer magnetoresistive RAM (STT-RAM) [23], ferroelectric RAM (FeRAM) [24], memristor [14], and nanomechanical memory (NEM) [6]. Typical performance parameters of mentioned memory technologies are presented in **Table 1** [14, 25].

Ferroelectric RAM (FeRAM or FRAM) is one promising memory, which employs a ferroelectric gate cell instead of a poly-silicon cell and has been considered as a replacement for flash memory [24]. In spite of some disadvantages like lower density, higher cost, and poor scalability, it has some advantages over flash like faster programming time, lower power usage, and higher endurance.

Magnetoresistive RAM (MRAM), which has been designed with magnetic storage elements, is another emerging technology [22]. The elements are made of two ferromagnetic

Traditional technologies					Emerging technologies				
			Improved flash						
	DRAM	SRAM	NOR	NAND	FeRAM	MRAM	PCRAM	Memristor	NEMS
Cell elements	1T1C	6T	1T		1T1C	1T1R	1T1R	1M	1T1N
Half pitch (F) (nm)	50	65	90	90	180	130	65	3–10	10
Smallest cell area (F2)	6	140	10	5	22	45	16	4	36
Read time (ns)	<1	<0.3	<10	<50	<45	<20	<60	<50	0
Write/erase time (ns)	<0.5	<0.3	105	106	10	20	60	<250	1ns(140ps-5ns)
Retention time (years)	Seconds	N/A	>10	>10	>10	>10	>10	>10	>10
Write op. voltage (V)	2.5	1	12	15	0.9-3.3	1.5	3	<3	<1
Read op. voltage (V)	1.8	1	2	2	0.9-3.3	1.5	3	<3	<1
Write endurance	10^{16}	10^{16}	10^{5}	10^{5}	10^{14}	10^{16}	10^{9}	10^{15}	10^{11}
Write energy (fJ/bit)	5	0.7	10	10	30	1.5 × 105	6 × 103	<50	<0.7
Density (Gbit/cm^2)	6.67	0.17	1.23	2.47	0.14	0.13	1.48	250	48
Voltage scaling	Fairly scalable					No	Poor	Promising	Promising
Highly scalable	Major technological barriers				Poor		Promising	Promising	Promising

Table 1. Traditional and emerging memory technologies characteristics [14, 25].

plates, each of them holds a magnetic field and an insulating layer separates them. One of the plates has a permanent magnetic field and the other has a variable one to be able to store data. MRAM is as fast as SRAM and as dense as DRAM. Also, it has a nonvolatility characteristic similar to flash and high endurance. However, it suffers from large cell size, high write current, and poor scalability, which greatly forbids being widely commercialized.

Another nonvolatile memory is phase change memory (PCRAM) [21]. Similar to optical storage devices, it stores data into a chalcogenide glass. The state of the glass is changed to crystalline or amorphous whenever an electric current passes through a heating element and

http://dx.doi.org/10.5772/105947

generates heat or quenches the glass. The main limitations of PCRAM are its high programming current and relatively long write/read time.

Memristor is considered to be one of the best candidates for future memory technologies. A memristor is a two-terminal nonvolatile memory and has been designed based on resistance switching effects [14]. Memristor has low write energy and high density due to multilayer crossbar architecture. Since the memristor crossbar-based architecture is highly scalable, it is predicted to be the selected candidate to use for future ultrahigh density memories. As no tunnel oxide is used in this device, memristor has higher endurance than flash memory. In spite of high density and endurance, read time is considerably high. Memristors may easily replace flash memories; however, they are not an appropriate option to employ in extremely fast system components.

Another promising candidate as a replacement for CMOS devices is the nanoelectromechanical switch (NEM) [6]. On/off state of NEM devices is determined by both electrical and mechanical forces between gate (a movable beam) and source terminals. Unique characteristics of NEM relays which are not available in conventional CMOS, such as zero leakage current and near infinite subthreshold slope, make them ideal for designing highly energy-efficient applications. In spite of low leakage current, NEM relays do not offer high turn-on current like CMOS transistors; moreover, they suffer from low endurance.

2.2. Nanoelectromechanical switches

Figure 1(a) and **(b)** shows the simplest NEM switch, 3T NEM that consists of three terminals: a cantilever beam (which is connected to the source terminal), a gate, and a drain. The voltage difference between the gate and the movable terminal (VGS) controls the position of the beam and state of operation. When VGS goes higher than a certain threshold value, called the pull-in voltage (Vpi), the electrostatic force exerted by the gate exceeds the elastic force of the beam and pulls the beam toward the drain until the beam collapses to the drain and forms a conductive channel from the beam (source) to the drain, thus closing the switch (on-state;

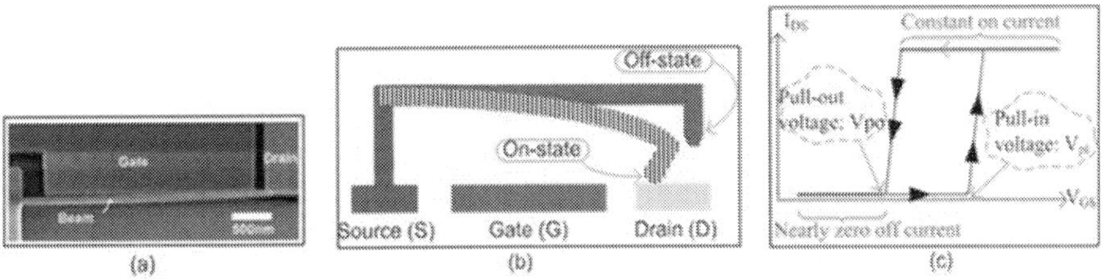

Figure 1. (a) SEM image of three-terminal NEM relay [28]. (b) Schematic of a 3T NEM relay consisting of source, gate, and drain; a cantilever beam connected to the source. The beam collapses to the drain when VGS ≥ Vpi and is released when VGS < Vpo. (c) Typical IDS-VGS characteristics of 3T NEM array.

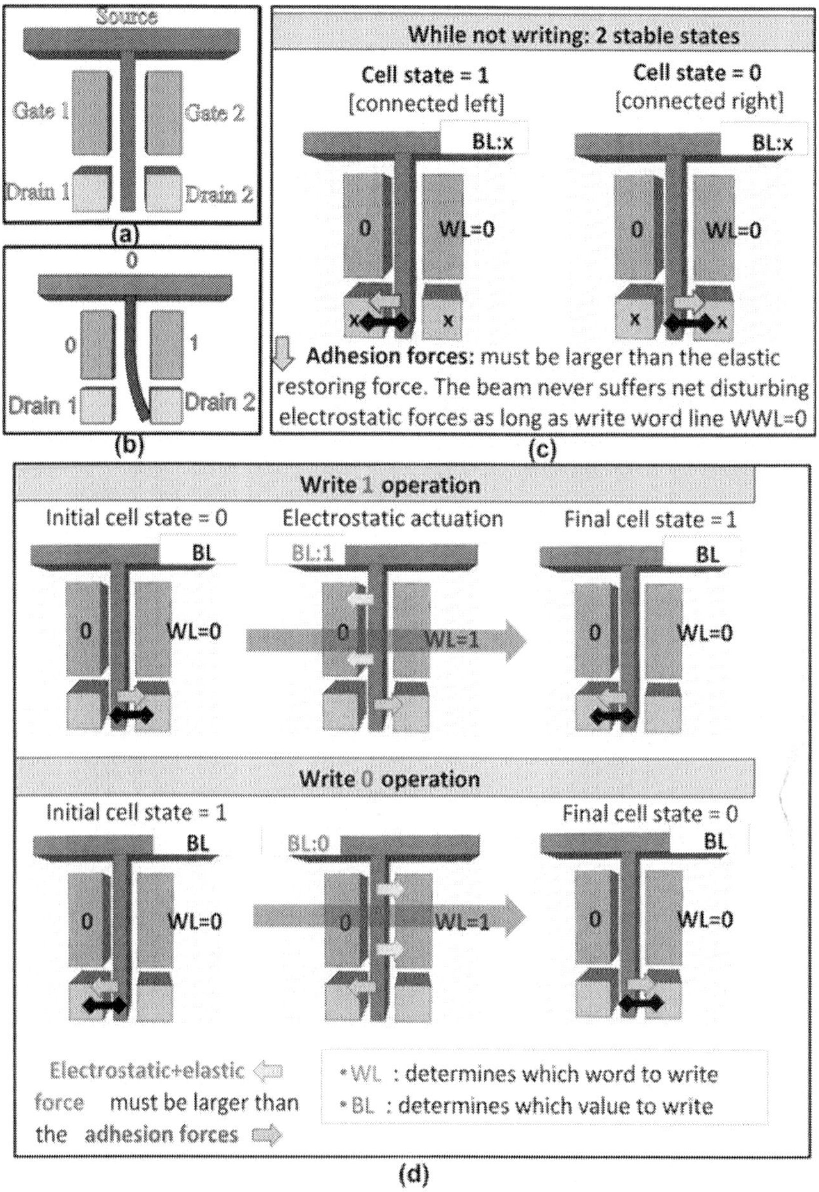

Figure 2. (a) A 5T NEM switch schematic. (b) The beam collapses to Drain2: VGS1 (VGate1-VSource) is 0 and VGS2 is 1. (c) The nonvolatile NEM considered here has two stable states. (d) Biasing scheme applied for writing the nonvolatile NEM switching. The BL assigns the value that will be written and chooses the cells for writing.

http://dx.doi.org/10.5772/105947

Figure 1c). In order to release the beam from the drain, VGS decreases to a voltage smaller than Vpi, called pull-out voltage (Vpo), where the electrostatic force is not higher than the elastic force of the beam; at this moment the beam is disconnected from the drain (off-state; **Figure 1c**). Due to such sharp on/off transition, NEMS have zero off-state leakage as there is no path for current to flow. Moreover, because of the surface adhesion force between the two contacting regions, Vpo is usually smaller than Vpi and the IV characteristics of NEMS exhibit such a hysteresis characteristic which enables NEM relays to be used as memory elements (**Figure 1c**).

There are several implementations of NEM switches [26, 27]. In this work, we consider 5-Terminal (5T) NEM switch which is illustrated in **Figure 2(a) and (b)**. A suspended beam is anchored at the source. Gate terminal 1 and gate terminal 2 (Gate1 and Gate2) are located close to the beam. The beam is connected to Drain1 or Drain2 (two output nodes). The beam moves toward Gate1/Gate2 because of the electrostatic attraction and a voltage difference between Gate1 and Gate2 (**Figure 2b**), and then connects to Drain1/Drain2, creating a conductive path between this drain and the source. An advantage of employing two gates is that the electrostatic force can be utilized as both pull-in and pull-out voltages. Therefore, one does not have to rely on only the elastic restoring force of the beam. Hence, the scalability of the device and the operational margins are improved considerably. However, the write operations in NEM switches take multiple clock cycles [11], because the mechanical movement of the beam is fairly slow which is related to the device technology.

2.3. Nonvolatile nanoelectromechanical switches

In this work, we choose the NEM switches which exhibit nonvolatile characteristic: once they are connected to a drain, they remain in this location until the beam is pulled out by electrostatic forces from the opposite gate. We select the NEM switch described in [29]. **Figure 2(c)** demonstrates the two stable states of this NEM switch. As long as both gate1 and gate2 are at the same potential (wordline, WL = 0), the beam never suffers a net disturbing electrostatic force. **Figure 2(d)** shows the write operation of this switch.

2.4. Memory arrays based on nanoelectromechanical switches

Former studies have evaluated employing NEMs for memory usages [11, 30]. Some of them address the use of NEM switches to replace normal memory arrays, such as SRAM. Chong et al. [11] replace the two pull-down transistors in a 6T SRAM cell with NEM switches to reduce leakage and area. Some of these studies also discuss nonvolatile memory arrays [30]. The memory array structure disclosed in Ref. [31] is of particular interest to our proposal as we explain in Section 3.

2.5. Configuration elements based on nanoelectromechanical switches

Recently, NEMs have been used for configuration tasks. Dong et al. [8] employed 3T NEMs as configuration memory components in FPGAs, replacing a routing switch by one NEMs, or an LUT cell by two NEMs. This design could be exploited for designing CAM cells; however, their proposed cell has many deficiencies. It relies only on the elastic restoring force for pull-out and suffers half-select conditions, it outputs only Out, not the complementary OutB, and it is volatile. The structure of the memory part of NEMsCAM which we describe in Section 3 depicts these mentioned shortcomings.

2.6. Content-addressable memory

A content-addressable memory (CAM) concurrently compares the search data with all of its stored data and returns the address of the matching location in a single clock cycle [27]. A typical CMOS-only CAM cell incorporates a SRAM cell to store the data bit and additional XOR circuits to compare the stored bit with the search data. CAMs propose a popular solution for a wide variety of applications that require high search speeds such as data compressions, network routers, and lookup tables [28]. However, the search operation in a CAM requires fully parallel comparison circuits to meet timing requirements. This results in high energy consumption and poses constraints on the number of entries affecting directly the effectiveness of the CAM.

2.7. Translation lookaside buffer

Virtual memory simplifies programming by abstracting and managing the available physical memory in pages. To accelerate virtual memory, processors employ the translation lookaside buffer (TLB) that holds recently used virtual-to-physical translations [20]. The processor searches the TLB on every memory operation using the virtual page number. In the case of a hit, the TLB returns the physical page number so that the memory operation can further proceed with accessing the memory hierarchy. However, in case of a miss, the memory operation will not complete until the address translation is retrieved from the memory (page walk) which might take up to hundreds of cycles. The TLB is hence a crucial component for the performance of the processor [20].

3. Design of NEMsCAM cell

In this section, we present the circuit details of our proposed NEMsCAM cell. We use the memory structure proposed in [31] to implement the storage part of NEMsCAM. That memory structure provides full-select behavior which is necessary to build a CAM; it also employs electrostatic pull-in and pull-out and does not need a cell selector component in the write path. The nonvolatile memory is designed based on the NEM switch proposed in [29], which can eliminate net-disturbing electrostatic force. **Figure 3(a)** depicts the configuration of the NEMsCAM cell. Out and OutB, the outputs, are connected to the transistors of the comparison part. We select CMOS for the comparison part to beware the long delay of the NEMs that happens because of the beams' mechanical movement, and that would slow down the search operation. **Figure 4(a)** represents the schematic of our NEM memory cell when it is programming to state "1." **Figure 4(b)** shows its switch model and **Figure 4(c)** shows a simplified NEM Verilog-A model between BL (source), Out (drain), and WL nodes, which we employ

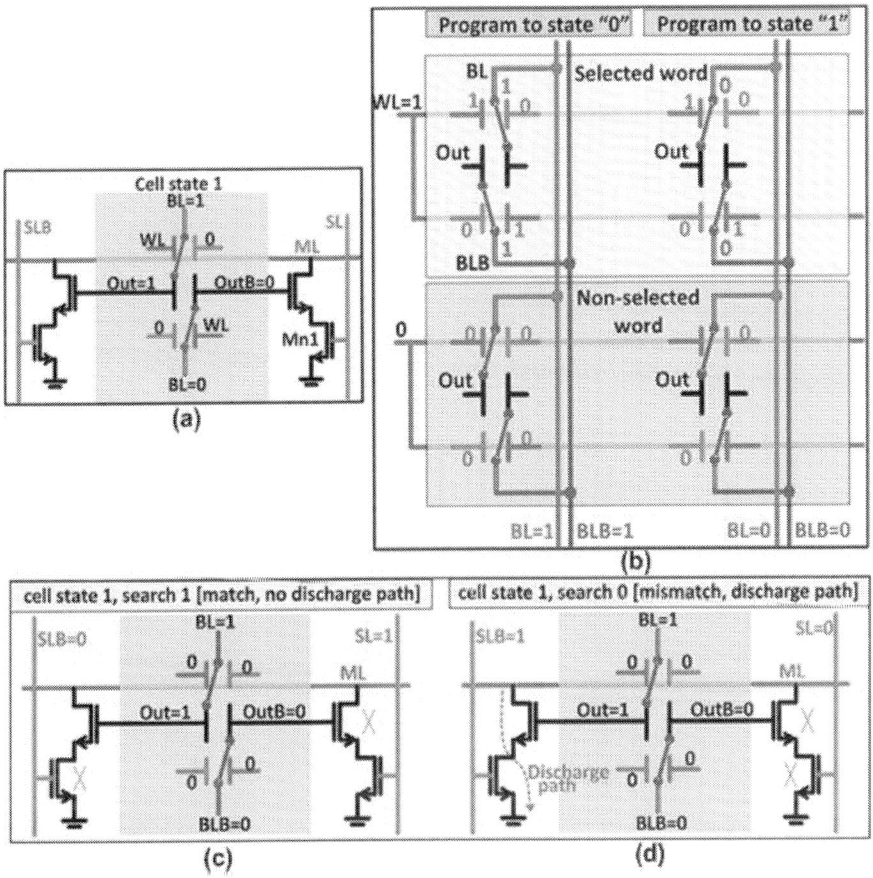

Figure 3. (a) Schematic of the proposed NEMsCAM cell. (b) NEMsCAM storage array organization and writing scheme. (c) When the NEMsCAM cell content matches the value being searched, there is no discharge path through the cell. (d) In the case of a mismatch, the cell tries to discharge the match line ML.

in our circuit analysis [11]. Other switches of NEM memory cell comply with similar switch modeling.

3.1. Circuit operations

3.1.1. Write biasing scheme

Figure 3(b) describes how the storage circuit of the NEMsCAM is written. When the WL (wordline) is activated, all beams on the row are sensitized. For the columns whose cells are to be programmed to 1, the bitlines are set to zero (BL = BLB = 0), and for the bitlines whose cells are to be programmed to 0, BL = BLB = 1 is applied.

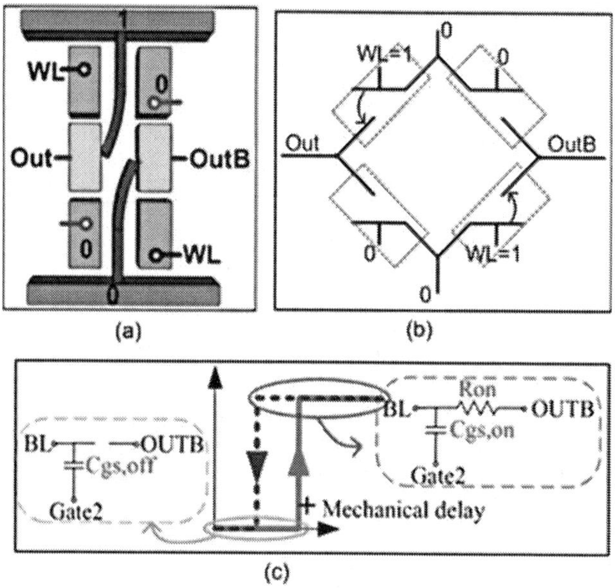

Figure 4. (a) Schematic of our NEM memory cell. (b) Simplified switch model of NEM memory cell. (c) A simple NEM Verilog-A model between BL (source), Out (drain), and WL nodes [11].

No cell suffers half-disturb situations, and since BLB and BL are always at the same potential during switching, there is no risk of short-circuit current running through the switches. This is critical because high currents through contact between the drain and the beam can be a source of failure. During typical operation, BLB is put at 0 and BL at 1. Cells whose beams are in state 0 hence have OutB = 1 and Out = 0. Keep in mind that there is no separate read operation in this memory design and there is no mechanical switch latency in the read path.

3.1.2. Search operation

Figure 3(c) and **(d)** shows a cell design that matches/mismatches the search data. If a cell(s) that is connected to a wordline entails a mismatch, the cell attempts to discharge the whole ML which is associated with that wordline, indicating an overall mismatch.

3.2. Cell architecture

Figure 5 presents the three-dimensional perspective of two neighbor NEMsCAM cells placed in the same column index of the array. As NEM switches have the possibility to be fully integrated with CMOS devices [32], they are located on top of the CMOS layer in this work and considerably decrease the layout space. The searchline (SL) wires are located parallel to the BL wires, whereas the matchlines (ML) and wordlines (WL) are located orthogonally to the BLs.

Using vertical NEM switches [27], the necessity of the long beam has a negligible effect on the layout space, since it is out of the plane. Two Gate1s are aligned and connected to their related WL, while the two Gate2s are connected to zero.

The drains are coupled from the opposite directions and build a cross configuration. The WL and BL wires can be combined with the real device terminals, resulting in a compressed layout. Eventually, the Vias connect OutB and Out to the CMOS layer which is placed under the NEMs layer. Because of this formation, our proposed NEMsCAM cell decreases the wire length, which considerably reduces the power consumption along with the near-zero leakage behavior of NEM switches.

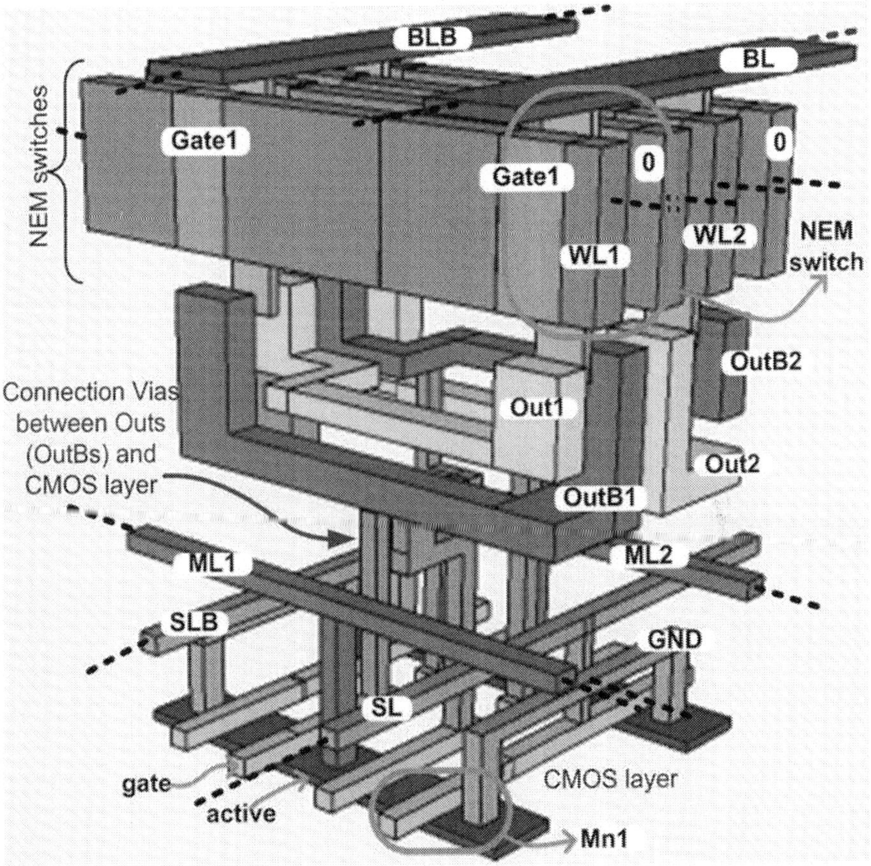

Figure 5. Three-dimensional view of two adjacent NEMsCAM cells in a CAM array.

4. A use case for NEMsCAM: TLB

As mentioned before, we leverage the design of the proposed NEMsCAM cell to build a fully associative translation lookaside buffer (TLB), called NEMsCAM TLB. In this section, we first elaborate on the motivation behind it and then we describe the design details and the circuit operations.

4.1. Motivation

Because of the importance of the TLB in the system's efficiency, processor designers have utilized a two-level TLB structure [33]. The first-level TLB is fully associative, small and provides a very fast search operation, while the second-level TLB is large and holds as many translations as possible. In order to achieve further system's performance, processors prepare separate TLBs for instructions and data [33].

The TLB hierarchy has been accounted for a substantial percentage of the power consumed in the chip [34, 35]. Intel recently informed that 13% of the total core energy comes from the TLBs designed for memory-intensive workloads [19]. Based on our evaluation base (Section 5), we discover that the TLB power consumption is overwhelmingly dominated by the first-level TLBs in terms of accesses across the TLB hierarchy (**Figure 6**). Moreover, by breaking down the power in the first-level TLBs, we detect that the CAM component contributes by 94%. In order to decrease this source of power consumption without diminishing the performance, we leverage our proposed NEMsCAM cell to design energy-efficient first-level TLBs.

4.2. Design

We design the NEMsCAM TLB with our proposed CAM cell and with typical SRAM memory circuits (**Figure 7**). The CAM part (**Figure 7a**) consists of the NEMsCAM cells and the neces-

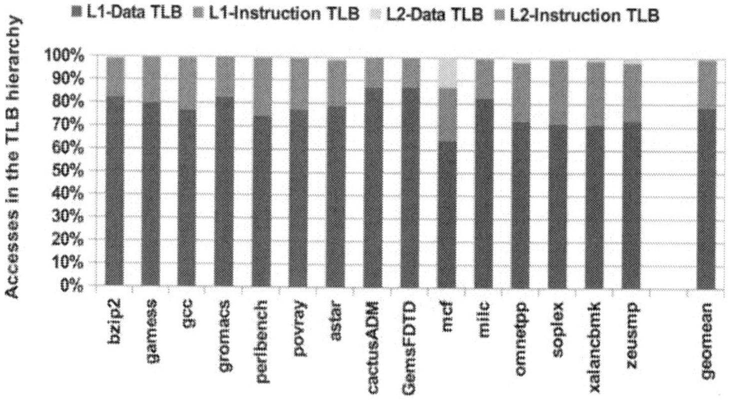

Figure 6. Breakdown of accesses in the TLB hierarchy.

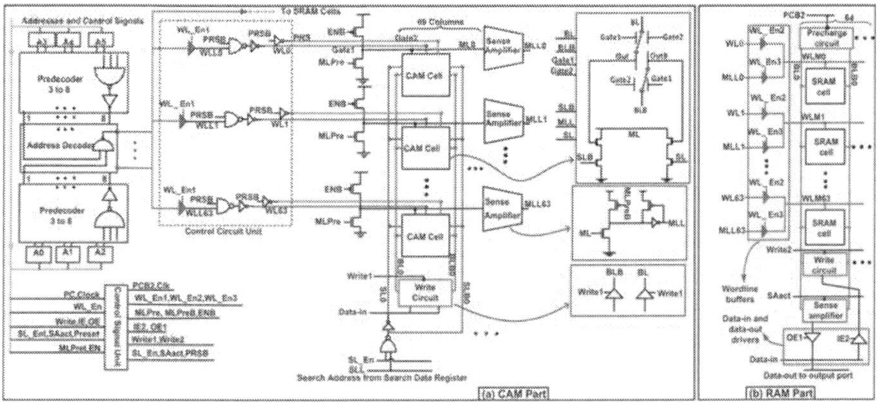

Figure 7. The circuit detail of (a) the proposed NEM-CMOS CAM and (b) a typical SRAM architecture in the proposed TLB structure.

sary peripheral circuitry optimized for both search and write operations. Similarly, the SRAM cells (**Figure 7b**) and the associated circuits are designed with CMOS technology. The control signal unit consists of the necessary inverter chains that generate the signals to control the TLB circuits so that the search and the write operations are performed correctly.

The address decoder, the write circuits, and the data-in drivers are used only for the write operation; however, the rest of the circuits are designed to be used during the search operation as well. BL and BLB are driven with predefined signals according to the operations. The control circuit unit is added to generate the necessary Gate1 and Gate2 signals during the search and write operations.

4.2.1. Search operation

Within the search operation, WLen3 becomes high and the WLM lines are connected to the ML lines. At the beginning of search operation, all ML lines are set provisionally in the precharged state as in a CMOS-only TLB. The search cycle begins when MLpre (the precharge signal) becomes high driving the ML to zero. At the same time, the SLs (search lines) are charged to their related data value; with this method, there is no need for a separate SL precharge phase. After this (completion of the precharge phase), the ENB signal becomes low and supplies the ML with the current source. During the evaluation phase, the stored bits of the CAM cells are compared against the data provided on the corresponding SLs.

In case of a match (TLB hit), the current source enabled by ENB pulls ML up and the ML voltage changes to high state. The state of each ML row is sensed and improved by an ML sense amplifier. Our used sense amplifier can be seen in this figure: an nMOS transistor, and also a half-latch circuit which stores the output data (**Figure 7a**). We choose the current-race scheme among various matchline-sensing techniques due to its simplicity and the average-low ML energy consumption [13]. Alternatively, in case of a mismatch (TLB miss) the cell(s) that cause a mismatch counteract the current source and keep ML close to ground level. In the match case, matchline trips the half-latch circuit when it is charged to a voltage slightly higher than threshold voltage of Msense; whereas, it leaves the latch in its initial state in the mismatch case when remains at a much lower voltage [5, 27]. Finally, the ML sense amplifiers feed the wordline buffers mapping the match location to its corresponding encoded address as stored in the SRAM cells (**Figure 7a**). **Figure 9** summarizes the signal behavior of the matching case for a cell of the NEMsCAM TLB.

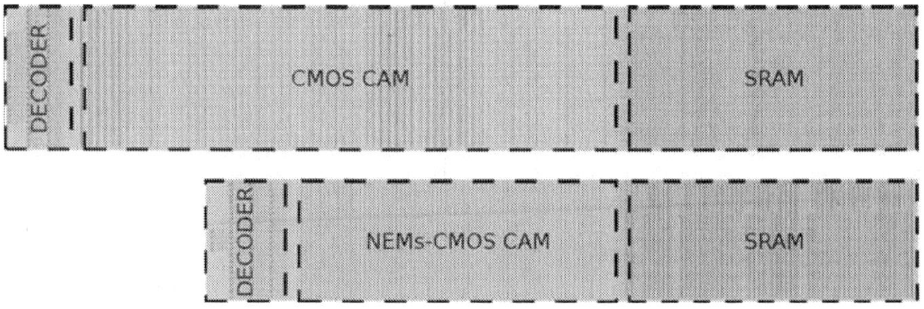

Figure 8. The layout of the DTLB implemented with CMOS-only CAM cells (top) and NEM-based CAM cells (bottom).

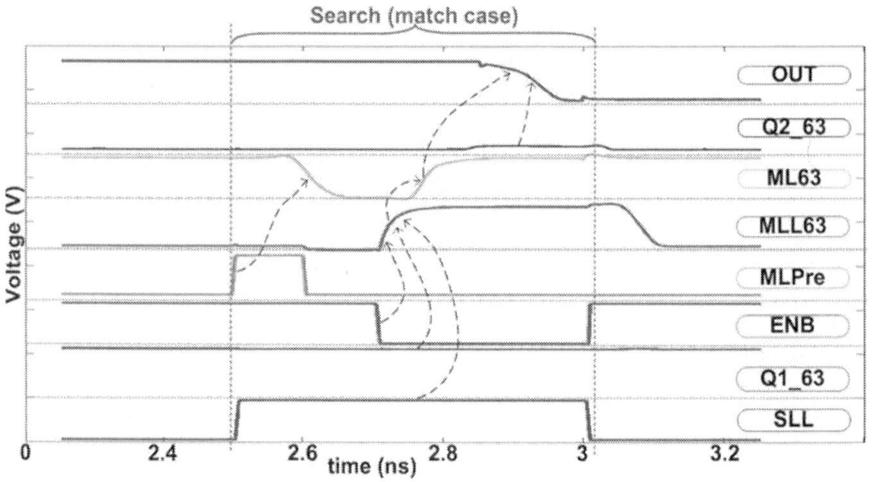

Figure 9. Simulation waveform of matching state for the first cell at the last row of the NEMsCAM DTLB.

4.2.2. Write operation

During the write operation, the WLen1 and WLen2 are high, the WL which is generated in address decoder is routed to the CAM and SRAM parts, and the data is written into the corresponding cells.

5. Experimental evaluation

In this section, we first describe our methodology to evaluate the NEMsCAM TLB, and then we present the results.

5.1. Methodology

We design NEMsCAM TLBs for DTLB (data) and ITLB (instruction) accesses based on [13] applying the TLB formation of a modern AMD server-oriented processor [33] (**Table 2**). For both CMOS-only and NEMsCAM TLB, we write the transistor-level netlists with all the tantamount resistance and capacitances of wires and essential circuitries. We simulate and optimize both TLB designs with Cadence Spectre exploiting 16 nm Predictive Technology Model [36] at $T = 25°C$ and 2 GHz processor frequency. As mentioned before, for the NEM switches, we apply a naive Verilog-A pattern (**Figure 4**) with the following parameters: Vpo = 0.2 V, Vpi = 0.8 V, Cgs-off = 15 aF, Cgs-on = 20 aF, tmech = 3 ns [11]. We optimize TLB circuits to minimize the energy consumption. We investigate that the search and write operations are performed correctly complying the timing necessities. We also assess the energy consumption per write and search operation and standby mode. Moreover, we plan the layouts [37], and span the wire lengths and optimize the wire capacitances in the netlists. Eventually, in order to evaluate the effect of the NEMsCAM TLBs at system's proficiency, we consume the Sniper simulator [38] with the configuration of **Table 2**, and run the TLB-intensive workloads from Spec2006 with the reference input set and execute for one billion instructions.

5.2. Results

5.2.1. Energy and area

Table 3 demonstrates the simulation outcomes for both CMOS-only and NEMsCAM TLBs. We perceive that the area decreases by 40.5% for the DTLB (**Figure 8**). The unique design of the NEMsCAM cell is the reason for this improvement. Furthermore, we discover that the energy consumption per write and search operation and standby mode decreases by 41.9%, 27%, and 53.7%, respectively, for the DTLB. This occurs due to the lower dimensions of the circuit leading to lower parasitic wire resistances and capacitances on the matchlines and the searchlines which in turn need fewer driving buffers. Also, the energy consumption further reduces due to the near-zero leakage current that NEMs prepare. Same results are achieved for the ITLB as well.

Per-core TLB organization		
Level 1	Data (DTLB)	64 entries, fully assoc.
	Instruction (ITLB)	48 entries, fully assoc.
Level 2	Data (L2-DTLB)	1024 entries, 4-way assoc.
	Instruction (L2-ITLB)	512 entries, 4-way assoc.

Table 2. TLB organization of a modern processor [33].

Structure	Metric	CMOS	NEMs	Benefit (%)
DTLB 64 entries	Search operation (pJ)	4.529	3.308	27.0
	Write operation (pJ)	0.148	0.086	41.9
	Standby mode (pJ)	0.141	0.065	53.7
	Area (normalized)	1.000	0.595	40.5
ITLB 48 entries	Search operation (pJ)	3.658	2.805	23.3
	Write operation (pJ)	0.187	0.107	42.8
	Standby mode (pJ)	0.106	0.046	55.9
	Area (normalized)	1.000	0.661	33.9

Table 3. Energy consumption for search, write operations, and standby mode and normalized area footprint.

5.2.2. *Latency*

Figure 9 demonstrates the simulation waveform of the matching case for one cell of the NEMsCAM DTLB (data TLB) during the search operation. The waveform considers that the design supplies the purpose time requirement of one clock cycle per search operation. On the other side, the write operation takes six cycles in the NEMsCAM TLB (based on Ref. [11]), whereas it takes two cycles in the CMOS-only TLB. This slowdown is because of the long mechanical delay of the NEM switches. However, this latency barely affects the processor performance as shown next.

5.2.3. *System*

Figure 10 displays the energy reduction in the first-level TLBs due to the NEMsCAM utilization for several workloads. We observe that the search operation overcomes in the energy breakdown for both ITLB and DTLB and that the NEMsCAM TLBs reduce the energy spent by 28.7% on average. Taking into account that 13% of the total core energy comes from the TLBs [19], the NEMsCAM cell can considerably assist in reducing the total chip's energy performance. **Figure 11** demonstrates the evaluated execution overhead due to the utilization of the NEMsCAM TLBs. This overhead occurs because of the increased latency of the write operation in NEMsCAM. Albeit, the write operation: (a) occurs only after TLB misses which take place scarcely compared to TLB hits, and (b) adds latency to an already slow operation,

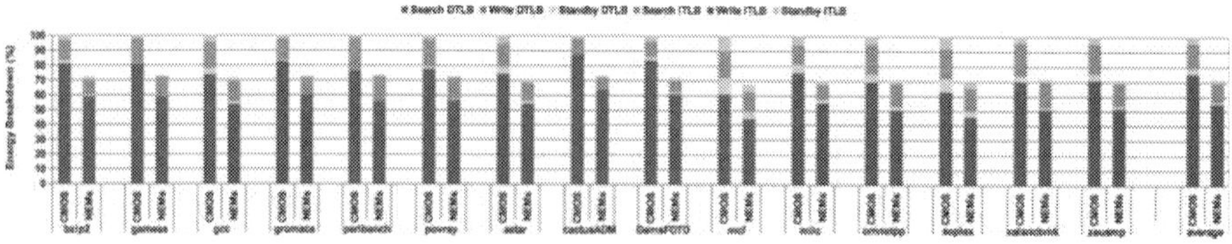

Figure 10. Dynamic energy consumption of the CMOS-only and the NEMsCAM DTLB and ITLB.

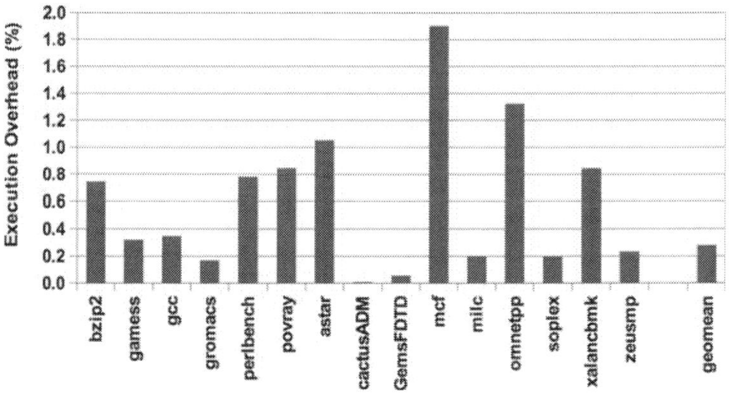

Figure 11. Execution time overhead due to NEMsCAM TLBs.

i.e., L2-TLB access (~7 cycles [39]), including potentially the penalty of L2-TLB miss (~100 s cycles [20]). Therefore, the NEMsCAMs TLB have an insignificant effect on the execution time for most applications (0.27% on average) while decreasing outstandingly the energy spent in the TLB hierarchy.

6. Summary

In this chapter, we describe the NEMsCAM cell design that combines both NEMs and CMOS to design low power and highly efficient processor structures such as TLBs. Our analysis shows that the NEMsCAM TLB exhibits significant benefits over the CMOS-only TLB in terms of energy consumption and area. However, the limited write endurance of current NEMs may delay their adoption until the technology improves.

Author details

Azam Seyedi[1,2]*, Vasileios Karakostas[1,2], Stefan Cosemans[3], Adrian Cristal[1,2], Mario Nemirovsky[1,4] and Osman Unsal[1]

*Address all correspondence to: azam.seyedi@bsc.es

1 Barcelona Supercomputing Center, Barcelona, Spain

2 Polytechnic University of Catalonia, Barcelona, Spain

3 IMEC International, INSITE, Leuven, Belgium

4 Catalan Institution for Research and Advanced Studies (CREA), Barcelona, Spain

References

[1] Moore G. E., Cramming More Components onto Integrated Circuits. Electronics. 1965; 38:114–7.

[2] International Technology Roadmap for Semiconductors, "Emerging Research Devices,2013 Edition." available on http://www.semiconductors.org/clientuploads/Research_Technology/ITRS/2013/2013PIDS.pdf

[3] Taur Y, Ning TH. Fundamentals of Modern VLSI Devices. 2nd ed. New York, NY, USA: Cambridge University Press; 2009.

[4] Dadgour HF, Banerjee K. Design and Analysis of Hybrid NEMS-CMOS Circuits for Ultra Low-Power Applications. In: 44th ACM/IEEE Design Automation Conference (DAC '07); 4–8 June 2007. p. 306–11.

[5] Akarvardar K, Elata D, Parsa R, Wan GC, Yoo K, Provine J, Peumans P, Howe R, Wong HSP. Design Considerations for Complementary Nanoelectromechanical Logic Gates. In: Electron Devices Meeting, 2007 IEDM 2007 IEEE International. 2007. p. 299–302.

[6] Nathanael R, Pott V, Kam H, Jeon J, Liu T-JK. 4-Terminal Relay Technology for Complementary Logic. In: IEEE International Electron Devices Meeting (IEDM). 2009. p. 223–6.

[7] Gaddi R, Schepens C, Smith C, Zambelli C, Chimenton A, Olivo P. Reliability and Performance Characterization of a Mems-Based Non-Volatile Switch. In: Reliability Physics Symposium (IRPS), 2011 IEEE International. 2011. p. 2G.2.1–2G.2.6.

[8] Dong C, Chen C, Mitra S, Chen D. Architecture and Performance Evaluation of 3D CMOS-NEM FPGA. In: Proceedings of the System Level Interconnect Prediction Workshop. Piscataway, NJ, USA: IEEE Press; 2011. p. 2:1–2:8. (SLIP '11).

[9] Chen F, Kam H, Markovic D, Liu TJK, Stojanovic V, Alon E. Integrated Circuit Design with NEM Relays. In: Computer-Aided Design, 2008 ICCAD 2008 IEEE/ACM International Conference on. 2008. p. 750–7.

[10] Han J-W, Ahn J-H, Kim M-W, Yoon J-B, Choi Y-K. Monolithic Integration of NEMS-CMOS with a Fin Flip-flop Actuated Channel Transistor (FinFACT). In: Electron Devices Meeting (IEDM), 2009 IEEE International. 2009. p. 1–4.

[11] Chong S, Akarvardar K, Parsa R, Yoon JB, Howe RT, Mitra S, et al. Nanoelectromechanical (NEM) relays integrated with CMOS SRAM for improved stability and low leakage. In: Computer-Aided Design - Digest of Technical Papers, 2009 ICCAD 2009 IEEE/ACM International Conference on. 2009. p. 478–84.

[12] Seyedi A, Karakostas V, Cosemans S, Cristal A, Nemirovsky M, Unsal O. NEMsCAM: A Novel CAM Cell Based on Nano-Electro-Mechanical Switch and CMOS for Energy Efficient TLBs. In: 2015 IEEE/ACM International Symposium on Nanoscale Architectures (NANOARCH). 2015. p. 51–6.

[13] Pagiamtzis K, Sheikholeslami A. Content-Addressable Memory (CAM) Circuits and Architectures: A Tutorial and Survey. IEEE J Solid-State Circuits. 2006;41(3):712–27.

[14] Eshraghian K, Cho KR, Kavehei O, Kang SK, Abbott D, Kang SMS. Memristor MOS Content Addressable Memory (MCAM): Hybrid Architecture for Future High Performance Search Engines. IEEE Trans Very Large Scale Integr Syst. 2011;19(8):1407–17.

[15] Matsunaga S, Hanyu T. Quaternary 1T-2MTJ Cell Circuit for a High-Density and a High-Throughput Nonvolatile Bit-Serial CAM. In: 2012 42nd IEEE International Symposium on Multiple-Valued Logic (ISMVL). 2012. p. 98–103.

[16] Nebashi R, Sakimura N, Tsuji Y, Fukami S, Honjo H, Saito S, et al. A Content Addressable Memory Using Magnetic Domain Wall Motion Cells. In: 2011 Symposium on VLSI Circuits (VLSIC). 2011. p. 300–1.

[17] Rajendran B, Cheek RW, Lastras LA, Franceschini MM, Breitwisch MJ, Schrott AG, et al. Demonstration of CAM and TCAM Using Phase Change Devices. In: Memory Workshop (IMW), 2011 3rd IEEE International. 2011. p. 1–4.

[18] Wang W. Magnetic Content Addressable Memory Design for Wide Array Structure. Magn IEEE Trans. 2011 Oct;47(10):3864–7.

[19] A. S. Race to Exascale: Opportunities and Challenges. In: MICRO Keynote. 2011.

[20] Jacob B, Mudge T. Virtual Memory: Issues of Implementation. Computer (Long Beach Calif) [Internet]. Los Alamitos, CA, USA: IEEE Computer Society Press; 1998;31(6):33–43. Available from: http://dx.doi.org/10.1109/2.683005

[21] Lee B, Ipek E, Mutlu O, Burger D. Architecting Phase Change Memory as a Scalable DRAM Alternative. In: International Symposium on Computer Architecture (ISCA) [Internet]. 2009. Available from: http://research.microsoft.com/apps/pubs/default.aspx?id=79150

[22] Zhu JG. Magnetoresistive Random Access Memory: The Path to Competitiveness and Scalability. Proc IEEE. 2008 Nov;96(11):1786–98.

[23] Heidel DF, Marshall PW, Pellish JA, Rodbell KP, LaBel KA, Schwank JR, et al. Single-Event Upsets and Multiple-Bit Upsets on a 45 nm SOI SRAM. IEEE Trans Nucl Sci. 2009;56(6):3499–504.

[24] Burr GW, Kurdi BN, Scott JC, Lam CH, Gopalakrishnan K, Shenoy RS. Overview of Candidate Device Technologies for Storage-Class Memory. IBM J Res Dev. IBM. 2008;52(4.5):449–64.

[25] International Technology Roadmap for Semiconductors, "Process Integration, Devices, and Structures,2011 Edition." available on http://www.maltiel-consulting.com/ITRS_2011-Process-Integration-Devices-Structures.pdf

[26] Ionescu AM, Pott V, Fritschi R, Banerjee K, Declercq MJ, Renaud P, et al. Modeling and Design of a Low-Voltage SOI Suspended-Gate MOSFET (SG-MOSFET) With a Metal-

Over-Gate Architecture. In: Quality Electronic Design, 2002 Proceedings International Symposium on. 2002. p. 496–501.

[27] Parsa R, Lee WS, Shavezipur M, Provine J, Maboudian R, Mitra S, et al. Laterally Actuated Platinum-Coated Polysilicon NEM Relays. J Microelectromech. Syst. 2013;22(3):768–78.

[28] Vaddi R, Pott V, Chua GL, Lin JTM, Kim TT. Design and Scalability of a Memory Array Utilizing Anchor-Free Nanoelectromechanical Nonvolatile Memory Device. Electron Device Lett IEEE. 2012 Sep;33(9):1315–7.

[29] Cosemans S. Data Storage Cell and Memory Arrangement. Google Patents; 2015.

[30] Kim M-S, Jang WW, Lee J-M, Kim S-M, Yun E-J, Cho K-H, et al. NEMS Switch With 30 nm Thick Beam and 20 nm High Air Gap for High Density Non-Volatile Memory Applications. In: Semiconductor Device Research Symposium, 2007 International. 2007. p. 1–2.

[31] Van Kampen RP. Four-Terminal Multiple-Time Programmable Memory Bitcell and Array Architecture [Internet]. Google Patents; 2009. Available from: http://www.google.com/patents/US20090273962

[32] Chen C, Parsa R, Patil N, Chong S, Akarvardar K, Provine J, et al. Efficient FPGAs Using Nanoelectromechanical Relays. In: Proceedings of the 18th Annual ACM/SIGDA International Symposium on Field Programmable Gate Arrays [Internet]. New York, NY, USA: ACM; 2010. p. 273–82. (FPGA '10). Available from: http://doi.acm.org/10.1145/1723112.1723158

[33] Advance Micro Devices. Software Optimization Guide for AMD Family 15h Processors. Number 47414. In 2014.

[34] Intel Strongarm Processor. www.intel.com/design/pca/applicationsprocessors/1110 brf.htm. In.

[35] Sh-3 RISC Processor family. http://www.hitachi-eu.com/hel/ecg/products/micro/32bit/sh 3.html. In.

[36] "Predictive Technology Model," Available on http://ptm.asu.edu/.

[37] "The Electric VLSI Design System," Available on http://www.staticfreesoft.com.

[38] Carlson TE, Heirman W, Eeckhout L. Sniper: Exploring the Level of Abstraction for Scalable and Accurate Parallel Multi-Core Simulation. In: Proceedings of 2011 International Conference for High Performance Computing, Networking, Storage and Analysis. 2011. p. 52.

[39] Intel Corporation, IntelR 64 and IA-32 Architectures Optimization Reference Manual. Number 248966–026. In.

Chapter 5

Energy-Aware Software Engineering

Kerstin Eder and John P. Gallagher

Additional information is available at the end of the chapter

http://dx.doi.org/10.5772/65985

Abstract

A great deal of energy in Information and Communication Technology (ICT) systems can be wasted by software, regardless of how energy-efficient the underlying hardware is. To avoid such waste, programmers need to understand the energy consumption of programs during the development process rather than waiting to measure energy after deployment. Such understanding is hindered by the large conceptual gap from hardware, where energy is consumed, to high-level languages and programming abstractions. The approaches described in this chapter involve two main topics: energy modelling and energy analysis. The purpose of modelling is to attribute energy values to programming constructs, whether at the level of machine instructions, intermediate code or source code. Energy analysis involves inferring the energy consumption of a program from the program semantics along with an energy model. Finally, the chapter discusses how energy analysis and modelling techniques can be incorporated in software engineering tools, including existing compilers, to assist the energy-aware programmer to optimise the energy consumption of code.

Keywords: energy modelling, energy analysis, energy transparency, energy aware, software engineering

1. Introduction

Energy-aware software engineering concerns the use of tools and methods to allow energy consumption to be a first-class software design goal. A design goal could be, for instance, to meet stated energy targets such as battery lifetime or power-supply constraints for a given ICT application running on a given hardware platform or simply to optimise energy efficiency.

Very few programmers at present have much idea of how much energy their programs consume or which parts of a program use the most energy. Therefore energy-related design goals are usually not considered until the programs are deployed; at that point, if energy goals are not reached, it may result in very long expensive redevelopment cycles.

Although energy is ultimately consumed by physical processes in the hardware, the software controls the hardware and indeed typically causes a great deal of energy waste by the inefficient use of the hardware. This waste cannot be recovered by relying on the development of more energy-efficient hardware—increasing the energy efficiency of the software is an essential part of reducing overall energy consumption [1]. Energy awareness for software development thus requires an understanding of the implications for energy consumption of design decisions in the software. In short, there is a need for *energy transparency*: the ability of the software developer to "see" the program's energy consumption, ideally without actually executing and measuring it.

Chapter outline. Section 1 presents the background and motivation for energy-aware software engineering. Then the main scientific and technical foundations that support energy transparency are summarised. These are *energy modelling* and *static energy analysis*. Energy modelling (Section 2) concerns building models of software energy consumption at different levels of abstraction, attributing energy consumption at the hardware level to software constructs such as operations, instructions, statements, functions and procedures. Energy analysis (Section 3) concerns the estimation, using an energy model, of the energy that would be consumed when running a piece of software, without actually executing it. This estimate can be parameterised by the input data for the software or other contextual information.

Section 4 contains a summary of the typical sources of energy inefficiency that can be removed when the programmer has relevant information on energy consumption. Finally, Section 5 describes how software designers and developers can use energy transparency during the software engineering process and what kind of activities constitute "energy-aware software engineering." For example, the programmer can analyse the program to identify which part of the software consumes most energy or explore the effect on energy consumption of different algorithms and data structures.

In contrast to much work on energy efficiency of ICT, this chapter adopts a generic approach, not driven by any particular class of applications, platforms or programming languages. The topic is currently mainly studied in different application contexts, such as embedded systems, high-performance systems, mobile systems and so on rather than as a coherent set of techniques applicable to any software-based system.

2. Energy-aware software engineering and Green IT

Concern over the increasing energy consumption and general environmental impact of ICT systems is growing. As a part of this, there has been a growth of interest in the field of *Green IT* [2–5] since approximately 2010; for example, the conference series International Green

http://dx.doi.org/10.5772/65985

And Sustainable Computing Conference[1] started in 2011 and the IEEE technical area of green computing[2] was launched in 2010. The Energy Aware COmputing workshop series[3] was initiated in Bristol in 2011. More recently, dedicated workshops such as GREENS[4] and SMARTGREENS[5] have been launched.

Green IT covers energy aspects of the complete life cycle and context of ICT systems, including software and hardware, development energy costs, maintenance and deployment energy costs, cooling costs, the energy costs of communication infrastructure, raw materials and disposal costs and a host of other energy costs and environmental effects associated directly or indirectly with software systems.

Energy-aware software development is therefore only one aspect of Green IT; it is only concerned with the energy efficiency of software, that is, the energy costs directly attributable to how programs use the hardware during execution. The energy-aware software engineer cannot in general be aware of the whole Green IT field, which involves complex dependencies and tradeoffs and goes well beyond software engineering.

2.1. Environmental motivation

The energy consumed by ICT is growing both in absolute terms and as a proportion of the global energy consumption and thus plays an important role in meeting the targets of the Europe 2020 Agenda, which includes a goal to reduce greenhouse gas emissions by at least 20% compared to 1990 levels. Every device, from autonomous sensor systems operating at the milliwatt level to high-performance computing (HPC) systems and data centres requiring tens of megawatts for operation, consumes a certain amount of energy which results in the emission of CO_2.

As already pointed out, energy is consumed by hardware, but the software often causes a great deal of energy waste by inefficient use of the hardware. Increasing the energy efficiency of the software is at least as effective as development of more energy-efficient hardware. Furber remarks that "if you want an ultimate low-power system, then you have to worry about energy usage at every level in the system design" [1]. Furthermore, in many cases the energy efficiency of software has a direct positive effect on the efficiency of other energy-related aspects of systems. Obvious cases are cooling costs and battery costs—cooling requirements for data centres are directly related to the power dissipated by the computations, whereas for mobile systems, the number of battery replacements or recharges is similarly reduced if software is more energy efficient.

---

[1] http://igsc.eecs.wsu.edu/ (formerly International Green Computing Conference (IGCC))

[2] http://sameekhan.org/tagc/

[3] http://www.cs.bris.ac.uk/Research/eaco/

[4] http://greens.cs.vu.nl/

[5] http://www.smartgreens.org/

2.2. Strategic motivation

The energy efficiency of ICT systems plays a critical role in exploiting the massive amounts of information available in data centres and the full vision of the so-called Internet of Things. The power requirement of a data centre is typically measured in tens of megawatts, including cooling costs, while the Internet of Things generates increasing demand for a huge number of very low-power devices. The dream of "wireless sensors everywhere" is accompanied by the nightmare of battery replacement and disposal unless the energy requirements of software running on devices can be lowered to enable them to be powered by energy harvesters or RF power sources.

2.3. Development costs of energy-efficient software

In the current state of the art, development costs for energy-efficient systems are higher than for energy-wasteful systems due to the extra effort required to take energy consumption into account. This is a significant barrier to making energy efficiency a first-class design goal.

The motivations for research in energy-aware software development can thus be summarised as follows:

1. To lower the energy costs directly attributable to software execution, helping to reduce the environmental impact of ICT and to enable the next generation of ambient low-power devices.

2. To lower energy costs indirectly caused by software, such as the cost of cooling, power supplies, battery replacement and recharging.

3. To reduce the costs of the process of developing energy-efficient systems, by developing tools and techniques to assist the energy-aware developer.

3. Energy modelling

An energy model supporting software energy analysis associates energy consumption costs with basic program constructs such as source code blocks, basic blocks in the intermediate representation used during compilation or machine code instructions. In addition, other costs arising from the execution of a program may need to be considered, depending on the micro architectural features of the hardware; examples are costs associated with the memory hierarchy, such as the cost of a cache hit and miss or the cost of accessing on-chip and off-chip memory, and also costs associated with the processor pipeline, such as the cost of pipeline stalls. In addition, the cost of the processor being idle and the cost of processing multiple threads concurrently may also need to be considered.

An energy model, as understood in this chapter, is program independent. It captures the energy costs of basic software constructs in a given language executed on a given hardware platform. The model is used during program analysis (Section 3) to obtain information about the energy consumption of a given program.

The challenge in energy modelling for software energy analysis is in finding a good compromise between the accuracy of the model and the ease with which the information can be mapped onto software constructs. Regarding the former, model accuracy tends to be higher for models at the lower levels of abstraction, i.e. instruction-level energy models are typically more accurate than energy models at the intermediate representation of the compiler, and source code energy models are less accurate in comparison. However, understanding which source code lines or blocks consume most energy is much more useful to software developers looking to optimise their code for energy efficiency than knowing the energy consumed by the sequence of machine instructions issued by the compiler. The higher the level of abstraction at which the information is presented to the software developer, the easier it is for them to comprehend the impact of algorithms and coding on the energy consumed during program execution. Yet, taking measurements to characterise energy models is simplest and most accurate when performed at the lower levels of abstraction, where energy costs of low-level software constructs such as machine instructions can be determined directly.

3.1. Defining and constructing an energy model at ISA level

The instruction set architecture (ISA) is the interface between hardware and software. It defines the hardware architecture and its behaviour in terms of low-level programming constructs such as supported data types, machine instructions, registers, the memory architecture, any interrupt and exception handling as well as I/O operations. The ISA provides a practical level of abstraction for energy modelling, because it is possible to directly correlate the energy consumption of the hardware operations associated with instruction execution to low-level software constructs.

Energy modelling at ISA level dates back to 1994 when Tiwari et al. [6] first proposed a method to develop instruction-level power models for arbitrary processor architectures to estimate the power consumption caused by software. Such models could overcome the limitations of hardware design power analysis methods, which require access to gate-level design information including layout and tend to be impractically slow at producing results for system-level power analysis. Instruction-level power models, instead, are orders of magnitude faster at estimating the power consumption of embedded software and can achieve accuracy within 10% of what hardware design power analysis methods deliver. This is a worthwhile trade-off because software development involves numerous iterations during the coding phase, and rapid feedback of resource usage is critical for software developers to make energy-aware decisions.

The model in [36] captures the energy consumption directly associated with processing each instruction, obtained by measuring the average current drawn while executing a dedicated loop that only contains independent instances of the respective instruction to be profiled, multiplied by the supply voltage V_{CC} and further multiplied by the number and duration of the clock cycles required to execute the instruction. Variations in instruction base costs can be observed during measurements and are due to different operand values being used during execution. It was observed that different operand registers lead to negligible variation while using different immediate values, or different memory addresses lead to observable yet small variation of no more than 5% for the architectures analysed. Because the exact operand values for instructions are only known at runtime, the energy model associates a single base cost with each instruction, representing averaged values. This is a very important feature of a single instruction cost Tiwari-style energy model as it has implications on the safety of the bounds inferred by worst-case static analysis techniques. This will be discussed further in Section 2.3.

Instruction base costs intentionally do not include any extra costs arising from executing an instruction within the context of other instructions, i.e. the overheads of executing arbitrary instruction sequences. One such cost is associated with switching the circuit state from executing one instruction to executing the next, termed the circuit state overhead. It captures the extra energy consumed due to switching on buses, e.g. as a consequence of changing opcodes and operand values, and using different functional units within the processor. The circuit state overhead is determined for all pairs of instructions by measuring loops that contain alternating sequences of the two instructions per pair. While including circuit state overheads into the energy model improves the accuracy of the model, the variation observed for individual instruction pairs was very limited for the architectures considered. It may thus be sufficient to determine a constant circuit state overhead cost and to use that instead of profiling all instruction pairs.

The execution of instruction sequences may give rise to other costs beyond the cost of circuit state switching, depending on the micro architecture of the processor. Resource contention due to data dependencies between instructions may cause pipeline stalls. Thus, the cost of pipeline stalls needs to be determined together with the number of stall cycles. In addition, there may be costs associated with cache misses, which typically cause execution delays of varying durations, depending on whether the fetch is from other cache levels of main memory. Their energy consumption also needs to be accounted for in an energy model, potentially sourcing information from a cache model that can provide cache miss rates for a given program. Thus, while instruction-level energy modelling techniques can be very accurate for simple architectures, the presence of complex architectural and micro architectural features such as several layers of caches, pipelines, superscalar processing, speculative execution, etc. can make it very difficult to achieve acceptable levels of accuracy when modelling at the instruction level.

In Ref. [7], this energy model is used to derive the energy consumption of a program *Prog* by Eq. 1:

$$E_{Prog} = \sum_{i}(B_i \times N_i) + \sum_{i,j}(O_{i,j} \times N_{i,j}) + \sum_{k} E_k \tag{1}$$

According to Eq. 1, the energy consumption of *Prog*, namely, E_{Prog}, is calculated as the sum of three components: the base cost of instruction execution, the circuit state overhead and other inter-instruction effects. The first term in the sum in Eq. 1 represents the base cost, where B_i is the base cost of instruction *i* multiplied with the number of times *i* occurs in program Prog , N_i. The second term is the circuit state overhead, where $O_{i,j}$ represents the cost incurred by switching the circuit state of the processor from executing instruction *i* to executing instruction *j*. This is multiplied by the number of times instruction *i* is followed by instruction *j* during the execution of program *Prog*, namely $N_{i,j}$. Finally, the third term in the sum accounts for the cost of k inter-instruction effects that may impact on software-related energy consumption, e.g. cache misses or pipeline stalls that can be characterised using external cache models or models of the micro architecture of the processor.

Equation 1 shows clearly the relation between the energy model and the analysis of the program. The terms B_i, $O_{i,j}$ and E_k are obtained from the energy model and are program independent. The terms N_i and $N_{i,j}$ are obtained by the analysis of the program, either dynamically, by profiling and counting the number of times each instruction or pair of instructions is executed, or statically as will be seen in Section 3.

Recently, instruction-level energy models have been developed for modern processors such as the Intel Xeon Phi, a many integrated core architecture for high-performance computing, and the hardware multi-threaded XMOS XCore embedded microprocessor [8].

The XMOS XCore instruction-level energy model [9] is based on the original model by Tiwari et al. However, it redefines the notion of base cost to be the power dissipated while the processor is idle and uses individual instruction costs, scaled by the level of concurrency in the processor's pipeline as well as a constant overhead to account for circuit state switching between instructions. Model characterisation was performed using a measurement setup and instruction loops similar to those originally proposed in Ref. [6]. The individual instruction costs represent averages over measurements obtained from running loops with instructions using operands that were generated pseudo-randomly, constraining values to those valid for the respective instruction. Evaluation of this multi-threaded instruction-level model showed average error margins of less than 7%. However, the model was designed to be used for static energy consumption analysis, requiring static analysis techniques to determine the number of idle cycles and the level of concurrent thread activity, in addition to the standard instruction stream statistics.

In contrast, the Xeon Phi instruction-level energy model [10] relies on performance counter statistics that are obtained at runtime, rather than through static analysis at compile time. This model is designed to be used with software profiling tools to support energy-efficient software development. The model is built by characterising the energy per instruction (EPI) of selected instruction types using microbenchmarks executed on different processor configurations in terms of numbers of cores and threads per core. Instructions are classified in terms of op-code and operand locations, both of which influence the EPI. The energy consumption of a given workload can then be determined by multiplying the runtime instruction statistics with the respective EPIs. This model achieves an average error rate of less than 5%.

Energy modelling at the ISA level gives us the following benefits: energy costs can be assigned at the instruction level, which is the same level as is output by the compiler; there are strong correlations between instruction properties and energy consumption, for example, the number of operands used in the instruction; and machine instructions can be related back to the original programming statements written by the software developer, as well as to various intermediate representations.

The construction of an energy model at the ISA level also has to address several challenges. Measurements need to be taken to determine both the base cost for each instruction and also the circuit state overhead. To achieve this, instructions are placed into infinite loops, i.e. loops of single instructions to obtain that instruction's base costs or loops of alternating instructions to characterise pair-wise circuit state overhead. The average current is measured while the loop is executed. Care needs to be taken to ensure the loop runs for a sufficiently long time to minimise measurement errors due to loop overheads. However, typically not all instructions can be directly profiled, requiring indirect or statistical approaches to their characterisation.

In general, for a modern processor with hundreds of instructions, the characterisation of the entire ISA is a significant effort. To reduce the measurement effort, rather than determining a base cost for each instruction, it may be sufficient to group instructions into classes of similar energy cost and to determine a single instruction class base cost. Likewise, instead of measuring circuit state overheads for individual instruction pairs, a cost that represents switching between instruction classes or a single constant circuit state overhead may be sufficient.

In addition, other properties such as the cost of running multiple threads and the cost of idle periods must be determined for multi-threaded architectures, and communication costs must be considered for interacting multi-threaded programs running on multicore platforms.

3.2. Energy modelling at higher levels of software abstraction

Instruction-level energy models are useful due to the close, almost direct link of measurements to programming constructs within the ISA. Energy models at higher levels of abstraction, however, provide more intuitive feedback to software and tool developers, albeit at the cost of accuracy of the predictions.

Modelling at the level of the intermediate representation (IR) used by compilers can be a useful compromise between the accuracy of a lower-level (ISA) model and the high-level source code. Since the compiler is a natural place for optimisation, modelling and predicting the energy consumption at IR level could therefore enable energy-specific optimisations.

IR-level energy models have been built using two distinct techniques. One is based on statistical methods and the other on mapping a lower-level model, i.e. one at ISA level, up to the IR level at compile time; both have been developed for LLVM IR [11] in the context of the LLVM toolchain [12].

In Ref. [13], statistical analysis has been employed to build an energy model for LLVM IR for the purpose of fast and accurate early-stage prediction of the energy consumption of embedded software at the function level to enable compile-time energy optimisation. Modelling starts with instrumenting and compiling the source code of a large set of benchmarks into architecture-independent LLVM IR to extract block-level statistics capturing the structural features of the source code in terms of LLVM instructions. Profiling of the LLVM IR basic blocks is then performed on a host computer to capture their dynamic behaviour in terms of basic block execution counts. The final step then factors in the timing and energy consumption of the target architecture. Using native compilation, back-annotation techniques to associate LLVM IR instructions with target machine instructions and statistical analysis, the target machine-specific costs are associated with the architecture-independent LLVM IR. This requires a cost model of the instruction set for the respective target machine, which is assumed to be provided by the manufacturer. The resulting model can be used to estimate the energy consumption of code for the target hardware, based solely on its target-independent LLVM IR.

A mapping technique that lifts an energy model at ISA level to LLVM IR level is described in Ref.[14]. The energy characteristics of LLVM IR instructions are determined from the costs of the associated machine instructions based on a mapping that tracks which LLVM IR instructions created which machine instructions during the lowering phase of compilation, i.e. after optimisation passes.

The approach provides on-the-fly LLVM IR energy characterisation that takes into consideration the context of instructions since there is no program-independent mapping between ISA instructions and LLVM instructions. Strictly speaking, the energy model is still at the ISA level; a program-dependent mapping is used to obtain energy costs of LLVM IR instructions and blocks at compile time. The technique has been used for static energy consumption analysis at the LLVM IR level in Refs. [14, 15] and also in Ref. [16]. The accuracy of energy consumption estimation at LLVM IR level is typically within 1–3% of that achieved using ISA-level models. This indicates excellent potential for exploiting this energy transparency during code generation. In principle, the same mapping technique may be used to map the energy consumption of programs to even higher levels, such as source code blocks or functions.

An alternative approach to building a source-level energy model, used in [17] to obtain a source code energy model for Android code, is to identify basic energy-consuming operations from the source code and correlate them to energy costs by measuring energy consumption in a large number of benchmark programs and analysing the results using techniques based on regression analysis. The resulting energy model of the basic operations implicitly includes the effect of all the layers of the software stack down to the hardware, including compiled code, virtual machine and operating system layers. The approach is inherently approximate; nevertheless such an approach may be the only feasible one in cases where the software stack has many complex layers, rendering a mapping-based approach difficult or impossible.

3.3. The impact of data on the energy model

The classic instruction-level energy model as described in Section 2.1 does not capture the impact of data on software energy consumption. For instance, a single energy cost is assigned to each instruction or instruction type. In practice, however, energy consumption is dependent on the data being processed, and for simple processors, data can make a significant contribution to energy consumption. This is illustrated in **Figure 1**, which shows the dynamic power, in mW, for the single-cycle XMOS XCore bitwise AND instruction for all 65,536 combinations of 8-bit operands from 0 to 255. The colours in the "heatmap" range from dark blue, indicating low power, to dark red, indicating high power. In Ref. [18], 15% data-induced variation has been reported for the 8-bit AVR processor, while up to 1.7x data-dependent variation was observed for the 32-bit XMOS XCore in [19]. Variation of as much as 50% is reported in [19] for an ST20 32-bit microprocessor.

However, static resource consumption-bound analyses must provide bounds on resource consumption that are both safe and tight, i.e. sufficiently close to the actual values. A good example is worst-case execution time (WCET) analysis [20], where under-approximation is not acceptable, i.e. unsafe, and significant over-approximation is considered not useful. Thus, for bound analysis, models must support the derivation of safe and tight bounds. A key prerequisite to achieve this for WCET analysis is timing predictability within a system, which enables precise bounds to be established with acceptable effort, without sacrificing performance of the computation in the general case. In fact, the architecture of processors can significantly impact on the design of analysis tools and the properties of the analysis results these can deliver [21] in terms of safety and precision. This is equally important for static energy consumption analysis, i.e. predictable architectures enable precise models to be developed.

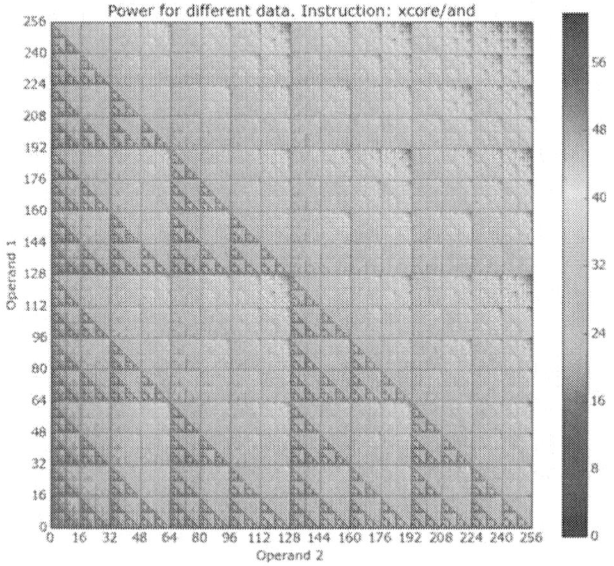

Figure 1. Dynamic power in milliwatts for the XMOS XCore AND instruction.

These observations naturally lead to the question of which energy cost should be associated with an instruction in a single-cost instruction-level energy model. Assigning the averages measured using randomly generated, valid data for the given instructions is a popular choice. Estimations based on such models may overpredict or underpredict the energy consumption of a program when compared to measurements. Error margins reported in the literature are typically below 10%, and overprediction and underprediction are both acceptable when the model is used to obtain estimations of energy consumption, and no guarantees are required.

It may be tempting to assign to an instruction the lowest or highest values observed during measurements to support best and worst-case analyses, respectively. However, this approach has been shown to lead to high over-approximation [22] in worst-case energy consumption static analysis, so this may not be a suitable option. In fact, the estimations based on such models may never be reachable in practice. Intuitively, this is because the data that causes the highest energy consumption for one instruction is very unlikely to produce output that will trigger the highest energy consumption also in subsequent instructions.

This leads us to the question of which input data causes the worst energy consumption for a given program. This question is investigated in [23], where the problem of data-dependent energy consumption during program execution is formalised in terms of circuit switching and a formal proof is presented demonstrating that in general analysing switching in processor datapaths is NP-hard. Thus, optimal data-sensitive worst-case energy consumption analysis of programs is, in general, not achievable efficiently, and alternative approaches giving good approximations must be developed. This is an area of ongoing research.

In Ref. [18], energy modelling for worst-case energy consumption analysis has been explored. The most promising approach uses probabilistic energy distributions to characterise individual instruction pairs and proposes techniques to compose these to block-level instruction sequences. In Ref. [19], activity indices were introduced into a single-cost instruction-level model to achieve higher precision of energy consumption predictions and to enable bound analysis for architectures where data significantly impacts on energy consumption.

4. Static analysis of energy consumption

Static analysis is the other key component of energy transparency. Given an energy model assigning energy costs to some basic units of the program, the task of analysis is to determine the overall energy consumption of the program or the distribution of energy consumption over the parts of the program. Static analysis infers information about energy consumed by programs without actually running them, in contrast to dynamic analysis, which collects information about the program's behaviour while executing it. Here we consider only static analysis.

As with energy modelling, analysis can be performed on program representations at different levels in the software stack, ranging from source code (in different programming languages) through intermediate compiler representations down to ISA level and employing an appropriate energy model at that level.

4.1. Semantic representations of programs

Static analysis of a program, and in general any formal treatment of programs, requires reference to a semantic model of the program derived from the semantics of the programming language in which it is written. Several different semantic styles and notations are used, including denotational semantic, small-step or structured operational semantics and big-step or natural operational semantics. All of these can be applied to code at various levels such as source code, intermediate compiler representations or ISA.

A common representation language, suitable mainly for operational semantics, is constrained Horn clauses (CHCs), a subset of first-order logic which is widely used in software verification [24]. CHCs can represent code semantics at any level of abstraction. In this section we outline the key aspects of resource analysis using CHCs as a representation, but space does not allow a fully detailed presentation. More information can be found in the references given in the text.

A constrained Horn clause has the form $\forall\, x_0 \ldots x_n(p_1(x_1) \wedge \ldots \wedge p_n(x_N) \wedge \phi \rightarrow p_0(x_0))$. When representing program semantics, each predicate $p_0, p_1, \ldots, p_n$ typically corresponds to a program point, and its respective arguments $x_0, x_1, \ldots, x_n$ are tuples representing the state before and/or after those points. A clause thus represents a relationship between program states, and the constraint ϕ expresses the relationship between the values of the state variables. A special case of a Horn clause is where $n \leq 1$, that is, there is at most one atomic formula on the left of the clause. Such a clause often represents a transition from the state at one program point to the next.

Figure 2 illustrates the use of Horn clauses to represent an imperative program in a C-like language (a). The constrained Horn clauses (b) represent a transition system (c) induced by the program's small-step operational semantics (the quantifiers in the Horn clauses are omitted). The predicates $r_1, \ldots, r_4$ represent the program points 1, …, 4 and $r_i(x_i)$ means that program point i is reachable with state x_i, where x_i is the tuple of variables in scope at that point.

```
1: void f(int n) {
       z = 1;
2:     while (n > 0) {
           z = z*n;
           n = n-1;
3:     }
       print(z);
4: }
```

(a)

$$true \rightarrow r_1(n)$$
$$(r_1(n) \wedge z = 1) \rightarrow r_2(n, z)$$
$$(r_2(n, z) \wedge n < 0 \wedge z' = z * n \wedge n' = n - 1) \rightarrow r_3(n', z')$$
$$(r_3(n', z') \wedge n = n' \wedge z = z') \rightarrow r_2(n, z)$$
$$(r_2(n, z) \wedge n \geq 0 \wedge print(z)) \rightarrow r_4$$

(b)

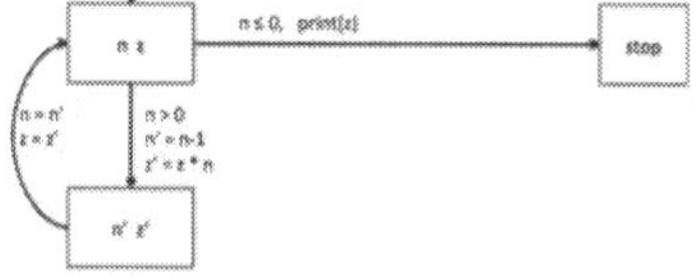

(c)

Figure 2. Transition system and constrained Horn clauses representing a program.

http://dx.doi.org/10.5772/65985

Lower-level programs such as ISA or intermediate code can be translated in a similar fashion, where typically each predicate represents a basic block in the code. Examples of the translation of XCore ISA programs to Horn clauses are given in Ref. [25]. Semantics-based methods for translating sequential imperative programs to Horn clauses are explained in Ref. [26]. Furthermore techniques for representing multi-threaded code as Horn clauses have been developed [27].

4.2. Techniques for energy analysis

Given such a representation of a program, techniques based on abstract interpretation [28] can derive safe approximations of program behaviour. In terms of CHCs, abstract interpretation can yield safe approximations of the values of the arguments of each predicate, which represent the set of possible states at some program point. A branch of abstract interpretation focuses on automatic complexity analysis, yielding complexity functions on the execution time of the program [29–33]. These techniques have been widely applied to analysis of Horn clauses and have been extended to analysis of energy and other resources [32, 33]. Tools, such as CiaoPP [34] and COSTA [35], have inbuilt facilities for resource analysis of programs including CHCs.

The essence of the techniques is to extract constraints from the Horn clauses representing the energy consumed. These constraints represent an abstraction of the behaviour of the program, in which the energy (or other resources being considered) can be considered as an implicit extra argument in the predicates of the Horn clauses (in some approaches, the extra-resource arguments are actually inserted into the Horn clauses, yielding a so-called "instrumented" representation). These constraints are then solved or approximated, to yield explicit formulas giving the consumption.

4.2.1. Linking analysis to an energy model

The Horn clauses in the semantic representation can be associated with energy values, using an energy model. For example, if the clauses are obtained from the source code, then each clause represents the execution of one statement or source code expression, and a corresponding source code energy model is associated with that clause. If the clauses are obtained from lower level code such as ISA, a clause typically represents the execution of an instruction or basic block; the corresponding energy consumption from the model can be mapped to the clause. The energy from a lower level model such as an ISA model can also be mapped to a Horn clause representing a higher level construct, possibly via an intermediate level as indicated in **Figure 3**.

Once this is done, constraints representing the energy consumption of the program are extracted from the Horn clause representation. To make the explanation more intuitive, we explain the process in terms of the transition system rather than the Horn clause representation. In the case of the loop at point 2, a recursive equation is obtained, e.g.:

$$cost_2(n) = e + cost_2(n-1)\ (\text{if } n > 0),\ \ cost_2(n) = 0\ (\text{if } n \leq 0) \tag{2}$$

where e is the energy cost of one iteration of the loop, obtained from the energy model. A dependency analysis also determines that the variable n is the relevant input parameter in this case. These equations can be solved to yield the expression giving the cost of the loop as a function of n, namely, $cost_2(n) = n * e$, and the cost of the whole program (a path from 1 to 4) is $cost(n) = e_1 + n * e + e_2$, where e_1, e_2 are the respective energy costs of the transitions before and after the loop.

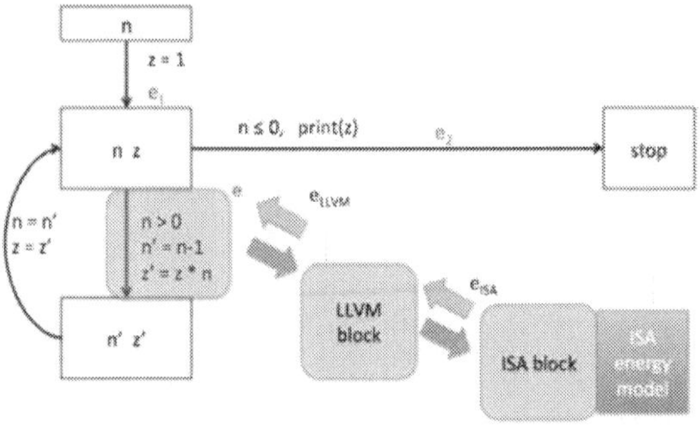

Figure 3. Combining an energy model with program analysis.

4.2.2. More complex analyses

The example shown is very simple, but the method generalises to more complex control and data structures. As the data and control-flow analysis of abstract interpretation is inherently approximate, the analysis in general gives safe upper and lower bounds on the number of times each part of the program is executed. This in turn gives upper and lower bounds on the energy consumed by the program. However, recall that the upper and lower bounds are also relative to the energy model, as discussed in Section 2.3. If the energy model supplies average costs for the basic instructions or operations, then the upper and lower bounds on the energy given by the analysis might not be safe, since the actual costs of executing the instructions might be respectively higher or lower than the average.

A further extension of the method of generating constraints and solving them yields static energy profiling [36], which shows the distribution of energy usage over the parts of the code, rather than a single function giving the total consumption of the program.

5. Software energy optimisation

One of the first works to stress the general importance of software energy efficiency, and identify aspects of software that affect energy consumption, was by Roy and Johnson [37]. More recent software-based approaches to achieving lower energy consumption are covered in [38, 39].

5.1. Computational efficiency

Firstly, there is a strong correlation between time and energy consumption for a given platform running a single computation thread. There are two reasons for this: less time means fewer instructions, and secondly when the task is finished, the processor can revert to a lower-power state for the excess time that a less efficient algorithm would use. The latter is called the "race to idle" in Ref. [39]. The correlation between time and energy is especially strong when asymptotic complexity is considered. It is highly likely, for example, that a single-threaded task that has $O(n^2)$ time consumption also has $O(n^2)$ energy consumption. Thus one of the main concerns of the energy-aware programmer, even with no knowledge of the energy consumption of the hardware, is to find computationally efficient algorithms and data structures suited to the task at hand.

5.2. Low-level or intermediate code optimisation

There is a range of techniques for low-level code energy optimisations, which could in principle be carried out by a compiler. These range from register allocation policies to avoid overheating a few intensively used registers, the use of very long instruction word (VLIW) instructions and vectorisation, to exploitation of low-power processor states using frequency and voltage scaling (DVFS). Note that such optimisations, in contrast to computational efficiency, are highly platform dependent and rely on a platform energy model expressed at the level of low-level code. Computational efficiency as described in Section 4.1 is also important in that low-level code optimisations are most effective when applied to frequently executed sections of code, such as tight inner loops, where a small saving in energy can make a significant difference to the overall computation.

Some energy optimisations rely on advanced compile-time (i.e. static) analysis. For example, knowledge of thread load imbalance and knowledge of predictable idle periods when processors can be put into low-power states are difficult to apply in the current compiler state of the art, since the analyses providing this knowledge are still emerging research areas.

5.3. Parallelism

The relationship between computational efficiency, time consumption and energy consumption is more complex for parallel than for sequential code. A multi-threaded solution using multiple cores can be more energy efficient than a single-threaded solution, even when the total amount of work done by the multi-threaded code is greater than that done by the single-threaded code. The savings are mainly due to the fact that the overall task time is reduced and so the processor(s) can revert sooner to a low-power state (the "race to idle" mentioned earlier).

Secondly, there can be energy savings if one or more cores can be run more slowly and still achieve the same overall task time as the sequential code. This is because power (P), frequency (f) and voltage (V) are related by the equation $P = cV^2f$ where c is a constant. Thus slowing down the processor (reducing f) saves power but not overall energy since the computation time is increased proportionally. However, a lower frequency is typically accompanied by a lower voltage, and the power/energy savings are quadratic in relation to voltage reduction.

5.4. Data and communication efficiency

Energy can be saved by minimising data movement. This can be achieved by writing software that reduces data movement using appropriate data structures, by understanding and exploiting the underlying system's memory hierarchy and by designing multi-threaded code that reduces the cost of communication among threads.

For example, the size of blocks read and written to memory and external storage can have a major impact on energy efficiency, while memory layout of compound data structures should match the intended usage in the algorithm, so that consecutively referenced data items are stored adjacently if possible. In multi-threaded code, consolidating all read-writes to or from disk to a single thread can reduce disk contention and consequent disk-head thrashing [39]. Furthermore, knowledge of the relative communication distances for inter-core communication can be used to place frequently communicating threads close to each other [40], thus reducing communication energy costs.

Synchronisation mechanisms can also have a serious impact on energy consumption. Waiting for events using polling loops is a notorious example as pointed out by Furber ("a polling loop [is] just burning power to do nothing") [1].

6. Software engineering activities and scenarios

We now look at energy-aware software from the designer's and developer's point of view. What are the activities that distinguish energy-aware design and development from standard approaches in which energy is considered at the end of the development process, if at all? In Section 5.1, we identify a number of generic activities that play an important role in energy-aware software engineering. In Section 5.2, we make the discussion a little more concrete by sketching scenarios in which these activities are applied.

6.1. Energy-aware software engineering activities

In this section we describe the most important activities involved in energy-aware software engineering. Some of these activities are extensions or modifications of conventional software engineering practices; others are new activities that only exist when energy efficiency is a design goal. **Figure 4** shows a number of activities and (some of) the interdependencies that arise in the context of different scenarios.

6.1.1. Specify application, including energy

The process of developing application software starts with a requirement specification that expresses not only *functional* properties, as in the classical approach, but also *non-functional* properties, including energy consumption and other resources. Classical methods for requirement specification need to be extended to allow non-functional specifications to be expressed.

Satisfying functional properties (in the sense of the classical concept of correctness with respect to a test suite or a formal input-output specification) is as important as doing so for non-

http://dx.doi.org/10.5772/65985

functional properties: an application that makes a device run out of batteries before a task is completed is as erroneous and useless as an application that does not compute the right result.

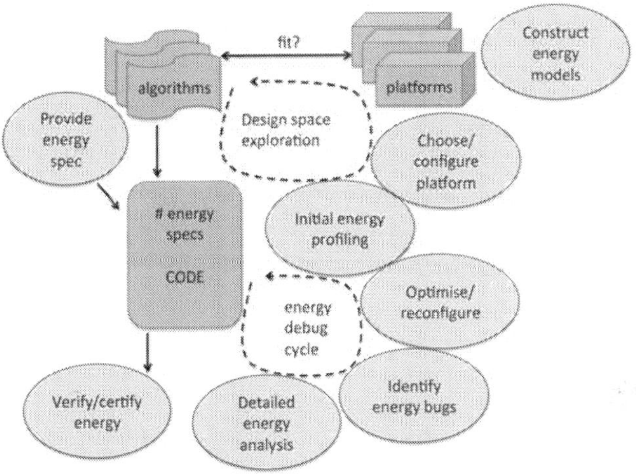

Figure 4. Energy-aware software engineering activities.

6.1.2. Construction of energy models

Creating energy models for different combinations of hardware platforms and programming languages is a part of the energy-aware development process. At one end of the spectrum, one might expect future hardware manufacturers to deliver an energy model for their instruction set architecture, and thus the model would be available "off the shelf." At the other end, some projects might require the construction of an energy model specific to that project, perhaps because the hardware or software environment was not standard. In between these two extremes, energy modelling for energy-aware software development is becoming a more well-understood process.

6.1.3. Resource model of deployment platform

If energy efficiency is a design goal, we need to obtain an *energy model of the platform* on which the system is to be deployed (even though the software might be developed on a different platform).

Thus, obtaining the appropriate energy model is a vital task in energy-aware software engineering. Not only should an appropriate platform be selected, but its energy model should be available during software development to support other activities (see, e.g., Sections 5.1.6–5.1.9). We note also that several different energy models for a given platform might be selected, at different levels of abstraction suitable for different activities. For instance, high-level approximate models might be suitable for design space exploration (Section 5.1.6) and initial energy profiling (Section 5.1.7), while more precise low-level models are needed for detailed energy analysis (Section 5.1.8) and optimisation (Section 5.1.10).

6.1.4. Selection of deployment platform

The choice of deployment platform itself might depend on its resource-usage model; thus, this activity and Section 5.1.3 are interdependent. By "platform" here is meant both the hardware and the software platforms; thus, the model should be capable of predicting the energy usage of software (in a given language and with a given runtime environment) being executed on a given piece of hardware.

6.1.5. Configure platform

Some platforms allow configuration that can have implications for energy consumption. Among such settings are clock frequency and voltage, the number of cores and the communication paths among them. At the software level, operating system settings can also be considered, such as the settings for power saving and the resolution of OS timer processes that can send interrupts to other processes.

6.1.6. Design space exploration

Choices taken early on in the design process can have a profound effect on the energy efficiency of the final result. *Design space exploration* as an energy-aware software development activity refers to the process of estimating energy implications of different possible design solutions, before they are implemented. It may involve especially activities such as selection of deployment platform (Section 5.1.4), platform configuration (Section 5.1.5) and initial energy profiling (Section 5.1.7). This involves energy modelling and analysis tools as in some other activities but with the difference that one is likely to be more satisfied with approximate models and thus rougher estimates of energy consumption rather than precise predictions.

6.1.7. Initial energy profiling

At early stages of energy-aware software design and implementation, tools are needed to perform an *initial energy analysis*. The purpose of this is to produce statically an *energy profile* that identifies the overall complexity of the energy consumption of the software and how energy consumption is distributed over the parts of the program. It could also at this stage identify energy bugs (parts of the application software that do not meet their energy consumption specification).

Initial energy analysis requires an *energy model* of the deployment platform at an appropriate level of abstraction. At early stages, parts of the software may be missing, and it might not be possible to compile it to machine instructions; thus, an approximate model based on a model of source code might have to suffice.

6.1.8. *Detailed energy analysis*

During more advanced stages of energy-aware software implementation, detailed energy analyses at finer levels of granularity are needed. These are provided by tools containing more precise low-level energy models of the platform, able to give precise estimates of the energy consumption of critical parts of the code, which could be targets for energy optimisation.

6.1.9. *Identify energy bugs*

Energy bugs occur when software does not conform to an energy specification. The specification might state some overall resource requirement in which energy consumption is implicit, for example, on the length of battery life. The bug in such a case could be some energy-consuming process that is more expensive than necessary, a service that is not switched off when required, threads that synchronise badly and spend too much time waiting and so on.

6.1.10. *Energy optimisation or reconfiguring*

The broad concept of energy optimisation is applied throughout the whole software engineering process and starts right at the beginning with design space exploration and selection of appropriate platform, algorithms and data structures.

The specific energy optimisation performed in this activity is driven by the detailed energy analysis and the energy model of the platform. Both manual and automatic optimisations can be applied; the energy analysis should point to the sections of code that use the most energy, either because they involve costly energy operations or because they are frequently executed (e.g., tight inner loops). This activity also includes application of energy-optimising compilers.

6.1.11. *Verify or certify energy consumption*

Energy-critical applications need to be certified with respect to an energy specification. Tools combining detailed energy models and precise energy analysis are required in order to compare the inferred energy consumption with the specification, either verifying conformance or certifying that it holds within some specified limits of behaviour such as input ranges.

6.2. Energy-aware software engineering scenarios

In this section, we sketch scenarios in which the activities described in the previous section are applied.

6.2.1. *Embedded system development on xCORE*

The ENTRA project[6] considered energy analysis of embedded systems implemented in the XC language and deployed on the xCORE multicore architecture. An energy-aware software development strategy for such applications involves the following energy-aware activities:

[6] http://entraproject.eu

- Energy specification by writing pragma comments in the XC source code. Such pragmas could express energy constraints derived from customer requirements on the power supply.
- Platform selection and configuration. The xCORE architecture is highly configurable both in terms of the number of cores and their interconnection. The choice and configuration are guided by an energy model applied to proposed solutions, taking into account thread communication energy costs in a given configuration, as described in more detail in [40].
- Detailed program-independent energy models of the platform at ISA level are available. Program-dependent energy models are obtained for XC and LLVM IR code for the application from the ISA model and used to perform more precise and detailed energy analysis of the application.
- Optimisations of expensive or frequently executed code is performed on the basis of the energy analysis.
- The energy-optimising compiler for XC is applied to the application.
- Pragmas in the code are verified using comparison of the energy consumption predicted by the analysis with the constraints in the specification.

6.2.2. Android app development

A case study on Android app energy optimisation was carried out [41]. The study involved energy modelling and optimisation of applications based on an established game engine. An energy specification was not given; the aim of the study was to use a source code-level energy model to identify the most energy-intensive parts of the code in a number of typical use cases and then apply manual optimisations, reducing energy usage directly and thus prolonging battery life.

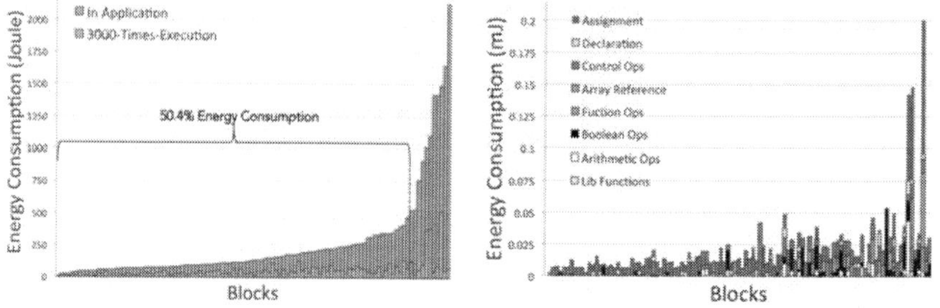

Figure 5. Energy distribution over basic blocks in an Android application. Blocks are sorted by the order of their contribution to runtime energy costs. The *green bars* indicate the relative costs of the blocks.

Energy-aware software engineering activities included:

- Building a fine-grained source code energy model by regression analysis from energy measurements on the target hardware and Android software platform of a set of test cases exercising the functions of the underlying game engine.
- Dynamic profiling of the code, which provided an energy profile that allowed the most energy-expensive basic blocks to be identified. For example, **Figure 5** from Ref. [41] shows an example of how the relative energy cost of basic source code blocks enable the programmer to focus optimisation effort on the most energy-consuming blocks.
- Manual refactoring of the source code, targeted at the most expensive blocks, which succeeded in increasing energy efficiency by a factor of 6–50% in various use-case scenarios.

7. Summary

The purpose of this chapter was to motivate energy-aware software engineering and to outline the principles and methods underlying it. We discussed why it is worth focusing on energy efficiency during software development and why energy efficiency should be a first-class software design goal.

A key concept is energy transparency, which makes the energy consumption of a program explicit at the level of code, rather than at the level of hardware, where the energy is actually consumed. A substantial part of the chapter described the two main fields of study relevant to energy transparency, namely, energy modelling and static analysis. Energy transparency is achieved by analysis of a program with respect to an energy model. The model associates energy consumption costs with basic units of the software, such as instructions or statements, and includes also other costs and overheads. Static analysis for energy consumption is a semantics-based formal technique, extending methods for automatic complexity analysis of programs, which is a branch of abstract interpretation.

The last part of the chapter considered the various features of software that can affect energy consumption. An energy-aware developer can use energy transparency to focus energy-aware design and optimisation in the most effective way. The field of energy-aware software engineering is only just emerging, and we described a number of activities that characterise energy-aware software engineering, extending or modifying conventional practices. The chapter concluded with a description of two short scenarios of energy-aware software engineering; however, a great deal of further experience and tool development is needed to realise the full vision.

Acknowledgements

This chapter is largely based on work done in the EU 7th Framework project ENTRA: Whole-Systems Energy Transparency (318337). Thanks are due to all the participants in the project, especially Pedro López García, Henk Muller, Steve Kerrison, Kyriakos Georgiou and Xueliang Li for material incorporated in the chapter.

Author details

Kerstin Eder[1] and John P. Gallagher[2*]

*Address all correspondence to: jpg@ruc.dk

1 University of Bristol, Bristol, United Kingdom

2 Roskilde University, Roskilde, Denmark

References

[1] S. Furber. Interview with Steve Furber: The Designer of the ARM Chip Shares Lessons on Energy-Efficient Computing. *ACM Queue*, 8(2), 2016.

[2] P.J. Krause, K. Craig-Wood, and N. Craig-Wood. Green ICT: Oxymoron, or Call to Innovation? In *Proceeding Green IT*, Singapore, 2010.

[3] S. Naumann, E. Kern, and M. Dick. Classifying Green Software Engineering–the GREENSOFT Model. *Softwaretechnik-Trends*, 33(2), 2013.

[4] E. Capra, C. Francalanci, and S. Slaughter. Is Software "Green"? Application Development Environments and Energy Efficiency in Open Source Applications. *Information & Software Technology*, 54(1):60–71, 2012.

[5] S.S. Mahmoud and I. Ahmad. A Green Model for Sustainable Software engineering. International Journal of Software Engineering and Its Applications, 7(4), July 2013.

[6] V. Tiwari, S. Malik, and A. Wolfe. Power Analysis of Embedded Software: A First Step Towards Software Power Minimization, pages 222–230. Kluwer Academic Publishers, 1994. 567021.

[7] V. Tiwari, S. Malik, A. Wolfe, and M. Tien-Chien Lee. Instruction Level Power Analysis and Optimization of Software. *The Journal of VLSI Signal Processing*, 13:223–238, 1996. 10.1007/BF01130407.

[8] XMOS. xcore: Architecture Overview. Technical report, XMOS Ltd., 2013.

[9] S. Kerrison and K. Eder. Energy Modeling of Software for a Hardware Multithreaded Embedded Microprocessor. *ACM Transactions on Embedded Computing Systems*, 14(3): 56, 2015.

[10] Y. Sophia Shao and D. Brooks. Energy Characterization and Instruction-Level Energy Model of Intel's Xeon Phi processor. In *International Symposium on Low Power Electronics and Design (ISLPED)*, pages 389–394. IEEE, November 2013.

[11] C. Lattner and V.S. Adve. LLVM: A Compilation Framework for Lifelong Program Analysis and Transformation. In *Proceeding of the 2004 International Symposium on Code Generation and Optimization* (*CGO*), pages 75–88. IEEE Computer Society, March 2004.

[12] LLVMorg. The LLVM Compiler Infrastructure, November 2014.

[13] C. Brandolese, S. Corbetta, and W. Fornaciari. Software Energy Estimation Based on Statistical Characterization of Intermediate Compilation Code. In *Low Power Electronics and Design (ISLPED) 2011 International Symposium on*, pages 333–338, Aug 2011.

[14] K. Georgiou, S. Kerrison, Z. Chamski and K. Eder. Energy Transparency for Deeply Embedded Programs. CoRR, abs/1609.02193, 2016.

[15] U. Liqat, K. Georgiou, S. Kerrison, P. Lopez-Garcia, M. V. Hermenegildo, J. P. Gallagher, and K. Eder. Inferring Parametric Energy Consumption Functions at Different Software Levels: ISA vs. LLVM IR. In M. Van Eekelen and U. Dal Lago, editors, *Foundational and Practical Aspects of Resource Analysis. Fourth International Workshop FOPARA 2015, Revised Selected Papers*, volume 9964 of Lecture Notes in Computer Science, pages 81-100. Springer, 2016.

[16] N. Grech, K. Georgiou, J. Pallister, S. Kerrison, J. Morse, and K. Eder. Static Analysis of Energy Consumption for LLVM IR Programs. In *Proceedings of the 18th International Workshop on Software and Compilers for Embedded Systems*, SCOPES 2015, pages 12–21, ACM. New York, NY, USA, 2015.

[17] X. Li and J. P. Gallagher. Fine-grained energy modeling for the source code of a mobile application. In T. Hara and H. Shigeno, editors, 13th Annual International Conference on Mobile and Ubiquitous Systems: Computing, Networking and Services (MobiQuitous 2016), pages 180-189. ACM. 2016.

[18] J. Pallister, S. Kerrison, J. Morse, and K. Eder. Data dependent energy modeling for worst case energy consumption analysis. CoRR, abs/1505.03374, 2015.

[19] G. Ascia, V. Catania, M. Palesi, and D. Sarta. An Instruction-Level Power Analysis Model with Data Dependency. *VLSI DESIGN*, 12(2):245–273, 2001.

[20] R. Wilhelm, J. Engblom, A. Ermedahl, N. Holsti, S. Thesing, D.B. Whalley, G. Bernat, C. Ferdinand, R. Heckmann, T. Mitra, F. Mueller, I. Puaut, P.P. Puschner, J. Staschulat, and P. Stenström. The Worst-Case Execution-Time Problem–Overview of Methods and Survey of Tools. *ACM Transactions on Embedded Computing Systems*, 7(3), 2008.

[21] R. Heckmann, M. Langenbach, S. Thesing, and R. Wilhelm. The Influence of Processor Architecture on the Design and the Results of Wcet Tools. *Proceedings of the IEEE*, 91(7): 1038–1054, July 2003.

[22] P. Wagemann, T. Distler, T. Honig, H. Janker, R. Kapitza, and W. Schroder-Preikschat. Worst-Case Energy Consumption Analysis for Energy-Constrained Embedded Systems. In *Real-Time Systems (ECRTS), 2015 27th Euromicro Conference on*, pages 105–114, July 2015.

[23] J. Morse, S. Kerrison, and K. Eder. On the infeasibility of analysing worst-case dynamic energy. CoRR, abs/1603.02580, 2016.

[24] N. Bjørner, A. Gurfinkel, K.L. McMillan, and A. Rybalchenko. Horn Clause Solvers for Program Verification. In L.D. Beklemishev, A. Blass, N. Dershowitz, B. Finkbeiner, and W. Schulte, editors, *Fields of Logic and Computation II–Essays Dedicated to Yuri Gurevich on the Occasion of His 75th Birthday*, volume 9300 of *Lecture Notes in Computer Science*, pages 24–51. Springer, 2015.

[25] U. Liqat, S. Kerrison, A. Serrano, K. Georgiou, P. Lopez-Garcia, N. Grech, M.V. Hermenegildo, and K. Eder. Energy Consumption Analysis of Programs based on XMOS ISA-level Models. In G. Gupta and R. Pea, editors, *Logic-Based Program Synthesis and Transformation, 23rd International Symposium, LOPSTR 2013, Revised Selected Papers*, volume 8901 of *Lecture Notes in Computer Science*, pages 72–90. Springer, 2014.

[26] E. De Angelis, F. Fioravanti, A. Pettorossi, and M. Proietti. Semantics-Based Generation of Verification Conditions by Program specialization. In M. Falaschi and E. Albert, editors, *Proceedings of the 17th International Symposium on Principles and Practice of Declarative Programming, Siena, Italy, July 14–16, 2015*, pages 91–102. ACM, 2015.

[27] S. Grebenshchikov, N.P. Lopes, C. Popeea, and A. Rybalchenko. Synthesizing Software Verifiers from Proof rules. In J. Vitek, H. Lin, and F. Tip, editors, *ACM SIGPLAN Conference on Programming Language Design and Implementation, PLDI'12*, pages 405–416. ACM, 2012.

[28] P. Cousot and R. Cousot. Abstract Interpretation: A Unified Lattice Model for Static Analysis of Programs by Construction or Approximation of Fixpoints. In R.M. Graham, M.A. Harrison, and R. Sethi, editors, *POPL*, pages 238–252. ACM, 1977.

[29] B. Wegbreit. Mechanical Program Analysis. *Communications of the ACM*, 18(9):528–539, 1975.

[30] M. Rosendahl. Automatic Complexity Analysis. In *4th ACM Conference on Functional Programming Languages and Computer Architecture (FPCA'89)*, pages 144–156. ACM Press, 1989.

[31] S.K. Debray and N.W. Lin. Cost Analysis of Logic Programs. *ACM Transactions on Programming Languages and Systems*, 15(5):826–875, November 1993.

[32] J. Navas, E. Mera, P. López-García, and M. Hermenegildo. User-Definable Resource Bounds Analysis for Logic Programs. In International Conference on Logic Programming (ICLP'07), Lecture Notes in Computer Science, pages 348–363. Springer, 2007.

[33] E. Albert, P. Arenas, S. Genaim, G. Puebla, and D. Zanardini. COSTA: A Cost and Termination Analyzer for Java Bytecode. In *Proceedings of the Workshop on Bytecode*

Semantics, Verification, Analysis and Transformation (BYTECODE'08), Electronic Notes in Theoretical Computer Science, Budapest, Hungary, Elsevier, April 2008.

[34] M. Hermenegildo, G. Puebla, F. Bueno, and P. Lopez-Garcia. Integrated Program Debugging, Verification, and Optimization Using Abstract Interpretation (and The Ciao System Preprocessor). *Science of Computer Programming*, 58(1–2):115–140, October 2005.

[35] E. Albert, P. Arenas, S. Genaim, G. Puebla, and D. Zanardini. COSTA: A Cost and Termination Analyzer for Java Bytecode. In Proceedings of the Workshop on Bytecode Semantics, Verification, Analysis and Transformation (BYTECODE'08), Electronic Notes in Theoretical Computer Science, Budapest, Hungary, Elsevier, April 2008.

[36] R. Haemmerlé, P. Lopez-Garcia, U. Liqat, M. Klemen, J.P. Gallagher, and M.V. Hermenegildo. A Transformational Approach to Parametric Accumulated-cost Static Profiling. In O. Kiselyov and A. King, editors, *13th International Symposium on Functional and Logic Programming (FLOPS 2016)*, volume 9613 of *LNCS*, pages 163–180. Springer, March 2016.

[37] K. Roy and M.C. Johnson. Software Design for Low Power. In W. Nebel and J.P. Mermet, editors, *Low Power Design in Deep Submicron Electronics*, volume 337, pages 433–460. Kluwer Academic, 1997.

[38] P. Larsson. Energy-Efficient Software Guidelines. Technical report, Intel Software Solutions Group, 2011.

[39] B. Steigerwald and A. Agrawal. Green software. In S. Murugesan and G.R. Gangadharan, editors, *Harnessing Green IT: Principles and Practices*, Chapter 3. John Wiley & Sons, Hoboken, NJ, USA, 2012.

[40] S.J. Hollis and S. Kerrison. Swallow: Building an Energy-Transparent Many-Core Embedded Real-Time System. In L. Fanucci and J. Teich, editors, 2016 Design, Automation & *Test in Europe*, pages 73-78, IEEE, March 2016.

[41] X. Li and J. P. Gallagher. A source-level energy optimization framework for mobile applications. In G. Bavota and M. Greiler, editors, *16th IEEE International Working Conference on Source Code Analysis and Manipulation (SCAM 2016)*, pages 31-40. IEEE Computer Society, 2016.

Chapter 6

Energy-Aware High Performance Computing

Martin Wlotzka, Vincent Heuveline, Manuel F. Dolz,
M. Reza Heidari, Thomas Ludwig,
A. Cristiano I. Malossi and Enrique S. Quintana-Orti

Additional information is available at the end of the chapter

http://dx.doi.org/10.5772/

Abstract

High performance computing centres consume substantial amounts of energy to power large-scale supercomputers and the necessary building and cooling infrastructure. Recently, considerable performance gains resulted predominantly from developments in multi-core, many-core and accelerator technology. Computing centres rapidly adopted this hardware to serve the increasing demand for computational power. However, further performance increases in large-scale computing systems are limited by the aggregate energy budget required to operate them. Power consumption has become a major cost factor for computing centres. Furthermore, energy consumption results in carbon dioxide emissions, a hazard for the environment and public health; and heat, which reduces the reliability and lifetime of hardware components. Energy efficiency is therefore crucial in high performance computing.

This chapter addresses key issues of energy-aware high performance computing. We outline some numerical methods which are often used in scientific applications, and present an energy profiling and tracing technique suitable to analyse the power consumption of applications. The next section is devoted to the performance and energy characterization of the sparse matrix-vector product, a basic numerical building block. Finally, we discuss opportunities for saving energy in computations by means of two examples. First, we present energy-aware runtimes on shared memory multi-core platforms for the Conjugate Gradient method. Second, we present energy-efficient techniques for multigrid methods on distributed memory clusters.

Keywords: high performance computing, energy-aware numerics, energy profiling, energy-aware runtimes, energy-efficient multigrid

1. Introduction

Numerical simulations play a key role in scientific discovery. Modeling and simulation of physical processes is an enabling technology for investigations beyond the scope of theoretical analysis and experiments. Numerical simulations may yield insight in situations where technical, ethical, or financial issues prevent from experiments. This chapter presents an overview of basic methodologies in high performance computing (HPC) used for numerical simulations of physical processes. We discuss energy-aware parallelization techniques both for shared memory multicore platforms, as well as for distributed memory clusters.

We outline a class of discretization methods that can be used to deal with physical models which are described by partial differential equations (PDEs) in Section 2. We demonstrate that solving algebraic systems of equations, and in particular solving linear systems, is at the heart of many simulations. In Section 3, we present techniques for power tracing and analysis of scientific applications. Section 4 is devoted to the energy characterization of the sparse matrix-vector multiplication. This is one of the fundamental numerical building blocks used in many solver methods. In Section 5, we focus on energy-aware runtimes for numerical linear algebra on multicore processors. Finally, we present a distributed memory parallelization of linear algebra operations and energy-efficient techniques for multigrid methods in Section 6.

2. Numerical simulation of physical processes modelled by partial differential equations

We depict the discretization methods and parallelization techniques by means of two prototypical partial differential equations, namely the Poisson's equation and the continuity equation. Poisson's equation can be used to model a steady state temperature distribution in a continuum, where the solution represents the equilibrium state. The time-dependent continuity equation can be used to model the temporal evolution of a conserved physical quantity. In our examples, we restrict to the case of two spatial dimensions. We outline the basic discretization methodology of the finite difference method (FDM) and of the finite element method (FEM) for Poisson's equation, and of the finite volume method (FVM) for the continuity equation. The goal is to show that these three methods have in common the translation of the underlying model equations into a finite-dimensional system of algebraic equations, either linear or nonlinear. The solution to the algebraic equations represents the numerical solution of the model. Thus, performing numerical simulations often requires to solve algebraic systems of equations. Therefore, the linear system solver is the crucial part of many simulations. This is also true for nonlinear systems, since these are often solved by means of Newton-type methods that use a sequence of linear systems to approximate the nonlinear solution.

2.1. Poisson's equation

The spatial domain where the model is defined is denoted by Ω. For our two-dimensional example, we have $\Omega \subset R^2$. Poisson's equation reads

$$-\Delta u = f \quad \text{in } \Omega, \tag{1}$$

where the unknown solution u represents the equilibrium temperature distribution in Ω, f is a heat source term, and $\Delta = \partial^2/\partial x^2 + \partial^2/\partial y^2$ denotes the two-dimensional Laplace operator. Poisson's equation is usually accompanied by boundary conditions. One can use Dirichlet-type boundary conditions to model given environmental temperatures, or Neumann-type boundary conditions to model given heat fluxes. Theory on existence and uniqueness of solutions can be found in textbooks, e.g., Refs. [1, 2].

2.2. Finite difference discretization of Poisson's equation

The finite difference method approximates derivatives by means of difference quotients, which are evaluated at certain points in the domain Ω. In the simplest case, the method is based on a rectangular grid Ω_h, covering the domain with squares of side length h which lie parallel to the coordinate axes. The grid points are denoted as $x_{i,j}$, where the indices i and j establish a lexicographic enumeration of points in the grid. In this sense, $x_{i-1,j}$ is the left neighbor of $x_{i,j}$ at a distance h and $x_{i,j+1}$ is the upper neighbor at a distance h. **Figure 1** shows an example domain with a grid. An approximation of the Laplace operator can be defined as

$$\Delta u(x_{i,j}) \approx \frac{1}{h^2}\left[u(x_{i-1,j}) + u(x_{i+1,j}) + u(x_{i,j-1}) + u(x_{i,j+1}) - 4u(x_{i,j})\right]. \tag{2}$$

Eq. (2) defines a discrete Laplace operator by means of the five-point stencil, which uses the function values at the point $x_{i,j}$ and its four neighbor points in direction of the coordinate axes.

Writing $u_{i,j} = u(x_{i,j})$ and $f_{i,j} = f(x_{i,j})$, this discretization of the Poisson equation results in the linear system of equations

$$4u_{i,j} - u_{i-1,j} - u_{i+1,j} - u_{i,j-1} - u_{i,j+1} = h^2 f_{i,j}. \tag{3}$$

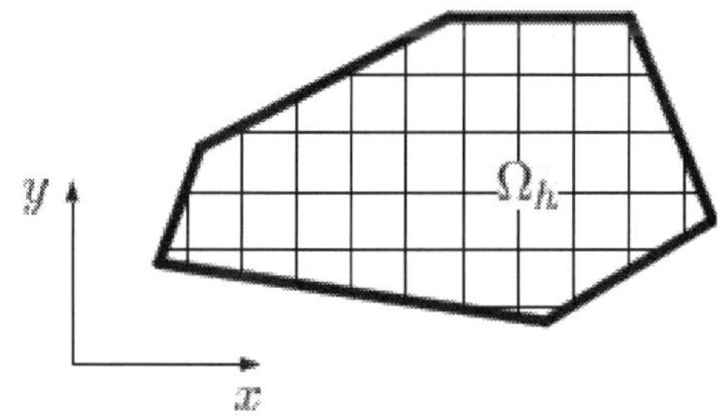

Figure 1. Simple computational grid Ω_h of a polygonal domain Ω.

Reusing the notation, the discretized system can be formulated with a matrix A and vectors u and b as $Au = b$. Here, the components of u and b represent the values of the discrete solution and of the source term at the grid points, respectively. The matrix A has the coefficients from the left-hand side of Eq. (3) as its entries. Thus, solving Poisson's equation by means of the finite difference method amounts to solving the linear system in Eq. (3) represented as $Au = b$. For error and stability analysis, and for convergence theory, see Ref. [3].

2.3. Finite element discretization of Poisson's equation

The finite element method uses a modified formulation of Poisson's equation. Multiplying Eq. (1) with a test function v and using Green's first identity, one obtains the variational formulation

$$a(u, v) = (f, v) \quad \forall \ v \tag{4}$$

with the bi-linear form $a(u, v) = \int_\Omega \nabla_v \cdot \nabla_u \, dx$, which admits a unique solution in the Sobolev space $H_0^1(\Omega)$ of weakly differentiable functions. Again, a rigorous derivation of the variational formulation, treatment of boundary conditions, and analysis can be found in Refs. [1, 2].

To approximate the solution of Eq. (4), the finite element method uses finite-dimensional function spaces of piecewise polynomials based on a grid Ω_h where we denote the vertices x_i. A simple case of such finite element function spaces of Lagrange-type is the space of piecewise first-order polynomials $V_h = \left\{ v_h \in H_0^1(\Omega) : v_h \Big|_C \in P_1 \text{ for all cells } C \in \Omega_h \right\}$ resulting in the problem formulation

$$u_h \in V_h : a(u_h, v_h) = (f, v_h) \quad \forall \ v_h \in V_h. \tag{5}$$

The Lagrange finite element basis functions have the property $\phi_i(x_j) = \delta_{ij}$ using the Kronecker symbol, meaning that $\phi_i(x_i) = 1$ and $\phi_i(x_j) = 0$ for $i \neq j$. Therefore, the support of any such basis function ϕ_i, i.e., the region where $\phi_i \neq 0$ comprises only the adjacent cells of the vertex x_i. Expanding the discrete solution $u_h = \sum_{i=1}^{n} u_i \phi_i$ in terms of an n-dimensional finite element basis with coefficients u_i, Eq. (5) is equivalent to the linear system of equations

$$\sum_{i=1}^{n} u_i a\left(\phi_i, \phi_j\right) = \left(f, \phi_j\right) \quad (j = 1, \ldots, n). \tag{6}$$

This can again be written as $Au = b$, where $A_{ij} = a(\phi_j, \phi_i)$ and $b_i = (f, \phi_i)$. For an overview of finite element theory, see Ref. [4].

2.4. Continuity equation

The prototype continuity equation reads $\partial_t \rho + \nabla \cdot j = f$, where ρ denotes the density of some conserved physical quantity, j is the flux, and f is a source term. The divergence operator applied to the flux is $\nabla \cdot j = \partial j / \partial_x + \partial j / \partial_y$. As an example for a conserved physical quantity, we consider the amount of some chemical substance which is dissolved in water with a concentration denoted as c. This concentration may vary in time and space. The flux j of the substance transported with a given water flow $\mathbf{v}$ is $j = c\mathbf{v}$. Thus, the equation of continuity for this example reads

$$\partial_t c + \nabla \cdot (c\mathbf{v}) = f \quad \text{in } \Omega \times (0, T). \tag{7}$$

The time interval under consideration is $(0, T)$, and Ω again denotes the physical domain where the model is defined. Note that we omit the discussion of initial and boundary conditions for simplicity here, although they are important for practical applications.

2.5. Finite volume discretization of the continuity equation

The finite volume method is well-suited to treat partial differential equations like the continuity equation [5]. Like the finite difference and the finite element method, it also uses a grid, Ω_h, covering the model domain. In this paragraph, we outline a simple cell-centered finite volume method. To this end, the grid cells $C_i \in \Omega_h$ are enumerated with the index i. Integrating Eq. (7) over a cell and using the divergence theorem yields $\partial_t \int_{C_i} c \; dx + \int_{\partial C_i} c\mathbf{v} \cdot \mathbf{n} \; ds = \int_{C_i} f \; dx$, where $\mathbf{n}$ denotes the outer unit normal vector field on the cell boundary. Denoting the average concentration of the solute in cell C_i as $u_i = \frac{1}{m_i} \int_{C_i} c \; dx$ with cell volume $m_i = \text{vol}(C_i)$, the boundary integral term can be approximated as

$$\int_{\partial C_i} cv \cdot \mathbf{n} \; ds = \sum_{j \neq i} A_{i,j} v_{i,j} u_{i,j}. \tag{8}$$

The summation index j runs over all cells adjacent to cell C_i, $v_{i,j}$ is the water velocity from cell C_i to cell C_j, $A_{i,j}$ is the cross-sectional area of the water flow between the two cells, and $u_{i,j}$ is the average concentration of the solute in the cell where the flow originates from. This leads to the spatially discretized system

$$\partial_t (Mu) + Au = b, \tag{9}$$

where the matrix M is a diagonal matrix with $M_{ii} = m_i$, the entries of the matrix A are determined through Eq. (8), and $b_i = \int_{C_i} f \; dx$. The resulting system of ordinary differential equation for the temporal evolution is often treated by means of numerical integrators such as one-step or multistep methods [6, 7]. When using implicit integrators, which is often the case for stiff problems, the computation of the time steps requires to solve linear systems of equations.

2.6. Commonalities

This brief outline on the basic methodology for discretizing partial differential equations by means of the finite difference, finite element, and finite volume method shows important commonalities. All three methods are based on a computational grid which covers the model domain. They transform the infinite-dimensional model problem into a finite-dimensional system of algebraic equations. Our linear example models directly yield linear systems. But also the solution of nonlinear equations is often approximated in Newton-type iteration with a sequence of linear systems. Thus, methods for solving linear systems of equations play the key role in many simulations. The structure of linear problems is induced by the association of the discrete variables with the grid cells or vertices. Couplings between the variables occur only locally between the neighboring grid cells or vertices. Therefore, the resulting discrete operators and matrices are sparse, i.e., there is only a small number of nonzero entries per row.

2.7. Numerical methods for solving linear systems of equations

As shown above, solving linear systems of equations is at the heart of many scientific applications. Such linear systems may directly arise from the discretization of partial differential equations, or when using a Newton-type iteration to approximate the solution of nonlinear systems. The numerical methods for solving linear systems can be classified in terms of being "direct" or "iterative". Direct linear solvers yield the solution after an a priori known number of computational steps. Prominent members of this class are the LU-, QR-, and Cholesky-factorization. In general, this class includes all methods derived from Gaussian elimination. Intermediate states of the solution vector in direct solver algorithms may be affected with arbitrarily large errors. Therefore, direct solvers must usually be executed entirely until the designated end of the algorithm. The computational complexity of direct methods is $O(n^3)$.

Iterative linear solvers compute a sequence of approximations $x^{(0)} \rightarrow x^{(1)} \rightarrow x^{(2)} \rightarrow \ldots$ which converges to the solution of the linear system. Iterative solvers are usually stopped after a certain number of steps and the recent iteration is taken as approximation to the solution. A typical stopping criterion is when the residual vector norm falls below a given tolerance.

Examples of iterative methods are the classic relaxation schemes such as Jacobi, Gauss-Seidel and successive over-relaxation (SOR), and derived relaxation schemes like asynchronous iteration. Widely used standard iterative solvers are the Krylov subspace methods such as conjugate gradient (CG), general minimal residual (GMRES) and their variants. In practice, Krylov subspace methods are often used with a preconditioner that accelerates the convergence of the sequence of approximations to the solution of the linear system. The computational complexity of Krylov subspace methods is $O(n^2)$ for sparse systems. Another type of iterative linear solver are multigrid methods. These are the most efficient state of art methods with a computational complexity of $O(n)$ for sparse systems. Iterative linear solver algorithms are built on a limited number of basic linear algebra operations. All iterative methods mentioned above use only vector scaling and addition, scalar product, vector norm computation, and matrix-vector multiplication. Therefore, any optimization of a linear algebra routine has a direct effect on the entire linear solver.

The applicability of any particular method depends on the mathematical properties of the system matrix. The matrix properties result from the underlying model equations and the discretization. Some methods such as the LU- and QR-decomposition or the GMRES method are applicable for general nonsingular matrices. Other methods require stronger properties like diagonal dominance or being symmetric positive definite. The latter is the case for the CG method. Textbooks on linear algebra and solvers include Refs. [8, 9].

3. Energy tracing and analysis for parallel scientific applications

The development of exascale systems has exposed the fact that the current technologies, programming practices, and performance metrics are not adequate. Existing tools for HPC systems mainly focus on monitoring and evaluation of performance metrics. Newer hardware has a wide range of sensors and measurement devices related to the power consumption, with varying granularity and informative value. Recently, new tools have been developed or have been adapted from other research areas (e.g., mobile computing) for analyzing the power consumption. Most of these tools focus only on the power consumption and disregard the performance aspects. Therefore, it is crucial to identify adequate sensors, hardware counters, and measurement devices to gain detailed insights about the power consumption and the performance of the hardware.

In order to optimize energy consumption of scientific applications, enhanced profiling and tracing frameworks combining both power and performance metrics are needed. Moreover, to gain a better understanding of energy usage, performance metrics, such as performance counters or routine events, should be correlated with the power traces. Only with analyzing these measurements, energy inefficiencies in software codes can be localized and optimized. In this chapter, we propose an integrated framework with a modular design to study power and energy profiles/traces of HPC scientific applications. This framework provides support for analyzing the power and performance of different types of parallel applications that run on distributed and shared memory platforms.

3.1. Framework environment

This section describes the software tools of the built-in framework that were developed for performance and energy profiling, and tracing of applications [10, 11]. This approach is based on the postmortem offline analysis as the recorded data is accessed after the application execution. The main advantage of this methodology is that the data can be analyzed many times and compared with other data.

The tracing mechanism normally proceeds out by collecting and analyzing data in order to characterize the application execution and system behavior. The approach to perform this analysis is realized in terms of statistics storage, comprising, e.g., absolute values for the number of invoked routines, the execution time of routines, and the hardware counters. Profiling tools that output these statistics are very useful to analyze the application behavior. Tracing tools are, as well, important to analyze the different phases and behavior of the application over time. Their extension to the power analysis also drives us to include data from the power measurement devices, with the aim of correlating them with the application traces. We use a combination of all these methods.

As shown in **Figure 3**, our proposed performance and energy/power analysis framework is composed of several components, including a power measurement library, a power measurement device, and a set of visualization tools. The scientific application runs on a high performance computing system (HPC), such as a cluster of multicore architecture or a hybrid system that benefits from offload computing parts such as GPGPUs. The application is instrumented with our power measurement library (PMLib), which allows for measuring the power consumption of the machine running the application. The second component is a measurement device which is attached to the target platform. This device is usually either an internal DC or an external AC wattmeter that steadily samples power and sends the output to a tracing server. The instrumented application makes calls to some API functions from the power measurement library for performing different and complementary tasks such as:

- Instructing the tracing server to start/stop collecting data captured by wattmeters.
- Dump the samples into a disk file (power trace) in a particular format.
- Querying various features of the measurement device, etc.

The last component of the framework is a visualization tool, which is used to analyze the power traces generated by PMLib, once the application finishes its computation. One can combine the power traces with the application performance traces produced by the Extrae tool to make the results more meaningful and visualize the final work with Paraver. Due to the flexibility of the framework and PMLib, it is also possible to take advantage of other tracing tools such as TAU [12], VampirTrace [13], etc.

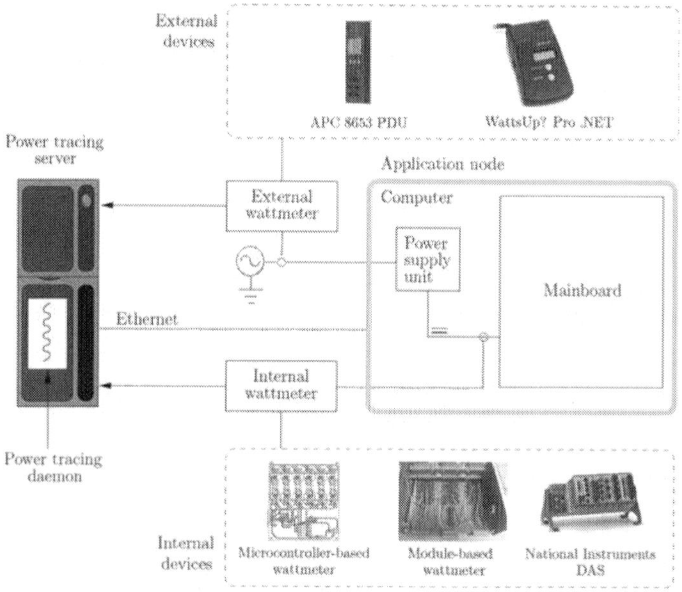

Figure 2. Single-node application system and sampling points for external and internal wattmeters.

3.2. The power measurement library: PMLib

The power measurement library (PMLib) has been designed to analyze the power consumption of HPC applications. The library supports various power measurement devices such as:

- AC wattmeters that connect to the input lines of power supply units of the computing systems and measure the AC input power. These devices include general power distribution units or professional wattmeters such as ZES Zimmer LMG450 or WattsUp?.

- DC wattmeters that can connect to the output DC lines of the power supply units inside the computing systems and measure the DC output power.
- Data acquisition cards such as National Instruments (NI) DAQs.
- Built-in power measurement components such as IPMI, Intel RAPL, and NVIDIA NVML.

The PMLib also offers a set of API interface to allow applications to measure their power consumption. The system and sampling points for external and internal wattmeters are illustrated in **Figure 2**.

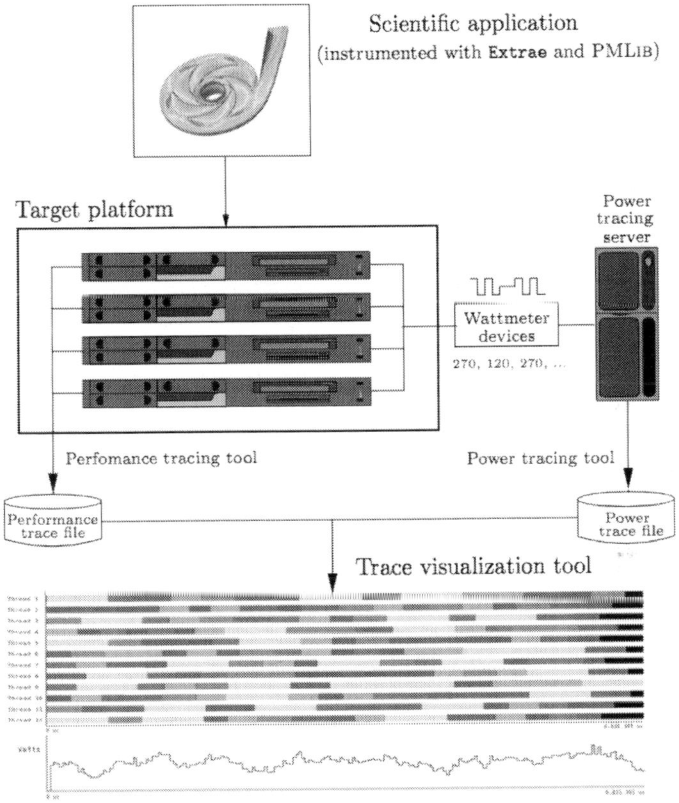

Figure 3. Collecting traces at runtime and visualisation of power-performance data.

3.3. Experimental results

In this section, we provide a detailed power and performance analysis of a dense linear algebra code to demonstrate the use of our performance-power framework on a multicore technology platform. This study offers a vision of the power drawn by the system during the execution of the LAPACK LU factorization [14]. This factorization is the key to the solution of dense linear systems. In order to collect and illustrate this information, we bind an execution trace of the algorithm obtained with our proposed framework Extrae + Paraver, with our own power evaluation setup, using PMLib and an internal DC wattmeter.

The following experiments were carried out using IEEE double-precision arithmetic on a two 8-core AMD Opteron 6128 processors, running at 2.00 GHz, with 48 GB of DDR3 RAM memory. This experiment benefits from the Intel MKL (v10.3.4) implementation of BLAS. Tracing and visualization were obtained with Extrae (v2.2.0) and Paraver (v4.1.0). In our evaluation, the power readings are collected from an internal DC wattmeter at a sampling rate of 28 Sa/s. The wattmeter is directly attached to the 12 V lines that connect the PSU to the motherboard (chipset plus processors) of the test platform. Therefore, the results are not affected by inefficiencies of the PSU, or the "noise" due to the operation of other hardware components such as fans, disks, network interfaces, etc.

The LAPACK implementation for the LU factorization with partial pivoting (dgetrf) was evaluated during our experiments. In this case, the parallelism is exploited within the invocations to the multithreaded MKL BLAS. The block size in LAPACK routines was set to 128 as it always delivers performance figures close to the optimal. **Figure 4** depicts

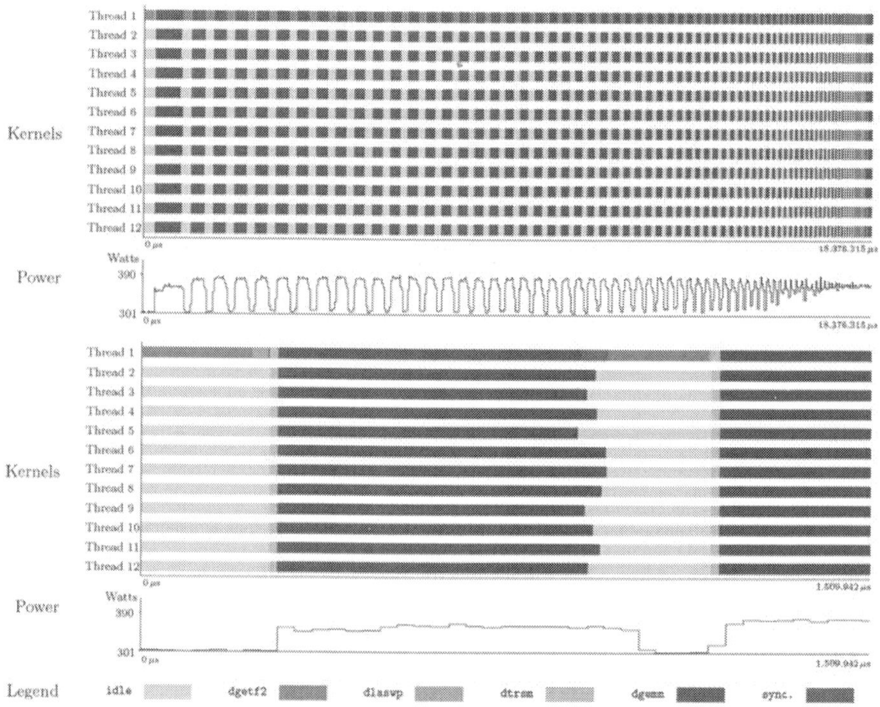

Figure 4. Trace of LAPACK dgetrf. Top: full. Bottom: first two iterations.

the activities of the cores and the power consumption traces during the execution of this routine, responsible for computing the LU factorization. In the trace, the colors identify the different kernels called by the main dgetrf routine: dgetf2 (factorization of the current panel), dlaswp (interchanging rows), dtrsm (triangular system solve), dgemm (matrix-matrix multiplications to calculate the trailing submatrix). The top trace in the figure shows the complete execution of dgetrf, while the bottom trace indicates the first two main iterations of the routine. As can be seen in the traces, the combination of this LAPACK routine along with the multithreaded MKL BLAS leads to interlaced sequential and parallel regions. It is also remarkable that the kernels, dgetf2 and dgemm, mostly monopolize the execution time of the routine. The zoomed trace in the bottom part reveals unbalanced loads among the execution units for the dgemm kernels, leading therefore to the need for the synchronization of threads. A closer look at the power traces highlights the fact that the recurring sequential and concurrent phases cause the power consumption to toggle repeatedly between 301 and 390W. For more details, refer to Ref. [15].

4. HPC energy metric and performance characterization of the sparse matrix-vector product

HPC metrics are supposed to drive system architects in the development of new hardware and supercomputers. With this respect, High performance Linpack (HPL) Linpack and the Top500 [16] have done a great job over the last 20 years, however with the advent of the power-wall and the imminent end of Moore's law, things started to change quite radically. Energy and power consumption are considered today as primary design parameters, together with FLOP/S and pure performance. For this reason, in 2007, the Green500 [17] ranking has been introduced, where FLOP/S have been normalized over power consumption. The new metric has finally moved some attention towards the power and energy consumption, however, the picture that it provides is still quite far from reality. This is visible in **Figure 5**, where FLOPS/W performance of the first system in Top500 and Green500 are compared between 2007 and today's. From the picture, it is clear that the 3.5x progress in last 5 years of Green500 has not propagated at all to the top systems in Top500. The motivation behind this behavior is that Green500 does not account for scalability and problem size. In other words, a small nonscalable system that barely enters Top500 could, in principle, rank first in Green500 due to a favorable MFLOPS/W ratio. Indeed, all top 10 systems in Green500 are rather small systems, with a consumption of O (100 kW), as shown in **Figure 6**.

The above scenario tells us that more has to be done at the level of the metrics to correctly capture real computational patterns. We need metrics that promote scalability rather than FLOPS. In other words, we need metrics that measure time- and energy-to-solution, with respect to problem size. This must also be accompanied by an increased typology of benchmarks, to capture all the variety of computations performed in real world applications.

To provide an example of the complexity and variety of the problem, in this chapter, we analyze the sparse matrix-vector product (SpMV). Sparse matrices appears in a lot of applications, with strong link to finite element models for partial differential equations, numerical methods for boundary value problems and also in economic modeling, ranking search methodologies for the web, and information retrieval. The sparse matrix-vector product (SpMV) has a central role in many of these applications [18], and is a key ingredient to address large-scale sparse linear systems and eigenvalue problems via iterative methods [8, 9]. Due to the relevance of SpMV in scientific computing, we pursue the accurate characterization of this operation on multithreaded architectures, from the point of view of three performance metrics, i.e., time, power, and energy. In this section, we present the most important aspects of Ref. [19].

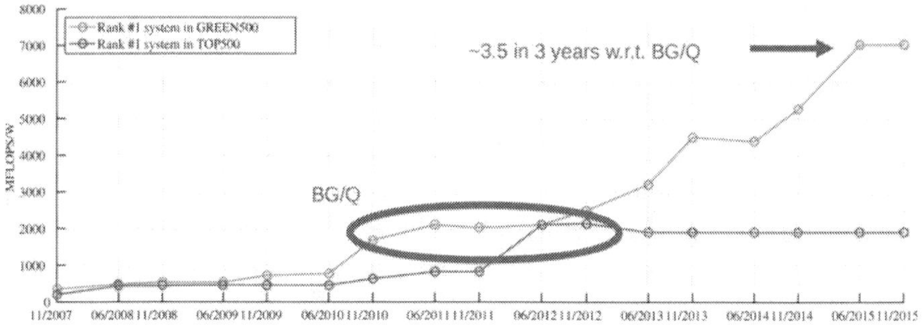

Figure 5. Comparison of the FLOPS/W performance of the no. 1 systems of the Top500 and the Green500 list.

Green500 Rank	MFLOPS/W	Site*	Computer*	Total Power (kW)
1	**7,031.58**	Institute of Physical and Chemical Research (RIKEN)	Shoubu - ExaScaler-1.4 80Brick, Xeon E5-2618Lv3 8C 2.3GHz, Infiniband FDR, PEZY-SC	50.32
2	**5,331.79**	GSIC Center, Tokyo Institute of Technology	TSUBAME-KFC/DL - LX 1U-4GPU/104Re-1G Cluster, Intel Xeon E5-2620v2 6C 2.1GHz, Infiniband FDR, NVIDIA Tesla K80	51.13
3	**5,271.81**	GSI Helmholtz Center	ASUS ESC4000 FDR/G2S, Intel Xeon E5-2690v2 10C 3GHz, Infiniband FDR, AMD FirePro S9150	57.15
4	**4,778.46**	Institute of Modern Physics (IMP), Chinese Academy of Sciences	Sugon Cluster W780I, Xeon E5-2640v3 8C 2.6GHz, Infiniband QDR, NVIDIA Tesla K80	65.00
5	**4,112.11**	Stanford Research Computing Center	XStream - Cray CS-Storm, Intel Xeon E5-2680v2 10C 2.8GHz, Infiniband FDR, Nvidia K80	190.00
6	**3,856.90**	IT Company	Inspur TS10000 HPC Server, Xeon E5-2620v3 6C 2.4GHz, 10G Ethernet, NVIDIA Tesla K40	58.00
7	**3,775.45**	Internet Service	Inspur TS10000 HPC Server, Intel Xeon E5-2620v2 6C 2.1GHz, 10G Ethernet, NVIDIA Tesla K40	110.00
8	**3,775.45**	Internet Service	Inspur TS10000 HPC Server, Intel Xeon E5-2620v2 6C 2.1GHz, 10G Ethernet, NVIDIA Tesla K40	110.00
9	**3,775.45**	Internet Service	Inspur TS10000 HPC Server, Intel Xeon E5-2620v2 6C 2.1GHz, 10G Ethernet, NVIDIA Tesla K40	110.00
10	**3,775.45**	Internet Service	Inspur TS10000 HPC Server, Intel Xeon E5-2620v2 6C 2.1GHz, 10G Ethernet, NVIDIA Tesla K40	110.00

Figure 6. Snapshot of the Green500 list.

4.1. Sparse matrices parameterisation

Let us consider the SpMV operation $y = Ax$, where A is an $n \times n$ sparse square matrix, with nz nonzero real elements, and x, y are both real dense vectors of dimension n. We assume that A is stored in compressed sparse row (CSR) format, using two integer vectors of dimension nz and $n + 1$ for the indices, and a real vector of size nz for the matrix entries [8, 9].

We do not limit our analysis to a specialized type of matrices (e.g., arrow, banded, tridiagonal, etc.) nor to a specific class of problems (e.g., matrices arising in electrical problems, structural problems, computational chemistry, ...). Our first goal is to analyze the problem and to identify a small set of parameters that capture the main "sparsity properties" of SpMV. A couple of these parameters, n and nz, are immediate and already appeared during the initial description of the problem. They are relevant because they determine the starting location of the matrix and vectors in the memory hierarchy that in turn has a big impact on the performance of SpMV. However, these quantities are not sufficient to describe in detail the sparsity pattern and they do not lead to an effective parameterization for our model. We thus introduce three complementary parameters (two of them normalized or nondimensional) to distinguish between matrices with similar dimension and number of nonzeros, but different sparsity patterns can later end in distinct characterization groups. In particular, for our CSR-based implementation of SpMV, we distinguish:

1. Block size: In many applications the nonzeros are clustered into a few compact dense blocks. The block size bs specifies the number of columns in these blocks. This parameter is important, because it captures the number of elements in vector x that are accessed with unit stride, which generally renders a better exploitation of data prefetching and the cache memory.

2. Block density: $bd = bs/nzr \in [0, 1]$ is the fraction of the nonzeros per row, nzr, occupied by a single block. This parameter is relevant because, with bs fixed, it characterizes the reuse factor in the access to vector y, i.e., for an average number of floating-point arithmetic operations performed each time, an element of vector y is loaded into a register (specifically, nzr FLOP).

3. Row density: $rd = nzr/n \in [0, 1]$ is the number of nonzeros per row relative to the row size. This ratio is orientational of the level in the memory hierarchy where the accesses to vector x occur. With bs, bd (and, consequently, nzr) fixed, increasing the row density rd necessarily implies a reduction in the problem dimension (and vice versa).

All these parameters will vary row-wise (bd and rd), and block-wise (bs). Several averaging techniques can be used to extract a single triplet (bs, bd, rd) from an entire nonuniform (irregular) sparse matrix.

4.2. Classification of matrices

We establish here a simple classification for SpMV (and thus for sparse matrices) with respect to the parameters previously defined, as well as according to performance metrics. We define a reference training set made of sparse matrices, with a uniform (though sparse) nonzero structure. In detail, the matrices in this set have a constant number of nonzeros per row *nzr* for all the rows, a fixed block size *bs* for all the blocks, and therefore the same number of blocks per row bd^{-1}, even though the position of the blocks inside each row is different and randomly assigned, with the only constraint that two blocks must be separated by at least one null entry.

To cover the full range of cases up to the memory capacity, while limiting the number of matrices in the set, we distribute the matrix instances equally in the $\log_2$ space:

- $bs = 2^0, 2^2, 2^4, \ldots$ up to 2^{12} on BG/Q and 2^{14} on P775,
- $bd = 2^0, 2^{-2}, 2^{-4}, \ldots$ down to 2^{-12} on BG/Q and 2^{-14} on P775,
- $rd = 2^0, 2^{-2}, 2^{-4}, \ldots$ down to 2^{-24} on BG/Q and 2^{-28} on P775.

Given that $bs \le nzr \le n$, we have a total of 118 samples on the IBM BG/Q and 175 samples on the IBM P755. **Figures 7** and **8** illustrate a compact 3-D representation of the training set, where each matrix is identified by a different point (*bs*, *bd*, *rd*) in the 3-D space.

This coarse training set gives enough variability to characterize sparse matrices from real applications. Nevertheless, we recognize that there exists a clear balance between training cost and accuracy.

We now classify the matrix instances of the training set into four groups, discriminating the training samples where the data fits into last level cache (LLC) from those that have to rely on DDR memory, while keeping the executions with one and four threads per core separated. Using a *k*-means clustering algorithm [20] (in the following we use $k = 2$), we establish a classification into *k* clusters per group for time, power, and energy. All measures are normalized a priori with the standard deviation because the algorithm relies on Euclidean distances.

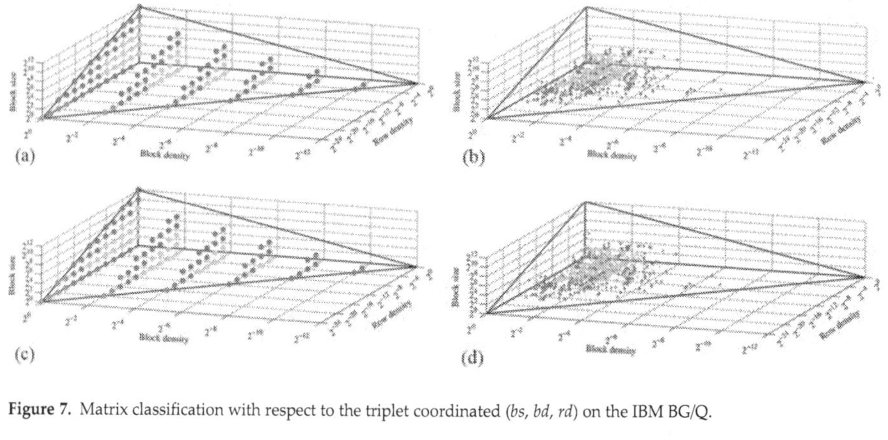

Figure 7. Matrix classification with respect to the triplet coordinated (*bs*, *bd*, *rd*) on the IBM BG/Q.

The left-hand side graphs in **Figures 9** and **11** illustrate the correlation between time and net power/energy. The resulting classification, summarized in **Table 1**, demonstrates that the *k*-means method is able to identify behavioral patterns in the time-power-energy triangle. At the same time, **Figures 7a** and **8a** show that the four classes are clustered into precise regions in the 3-D space defined by the triplet coordinates (*bs*, *bd*, *rd*). As we can see the classification holds with respect to the parameterization introduced in Section II and, in consequence, it can be leveraged to obtain a fast, qualitative prediction of low/medium/high time-power-energy behavior for any sparse matrix, as a function of few pattern information. In addition, we note that the same classification holds independently with respect to the number of threads per core.

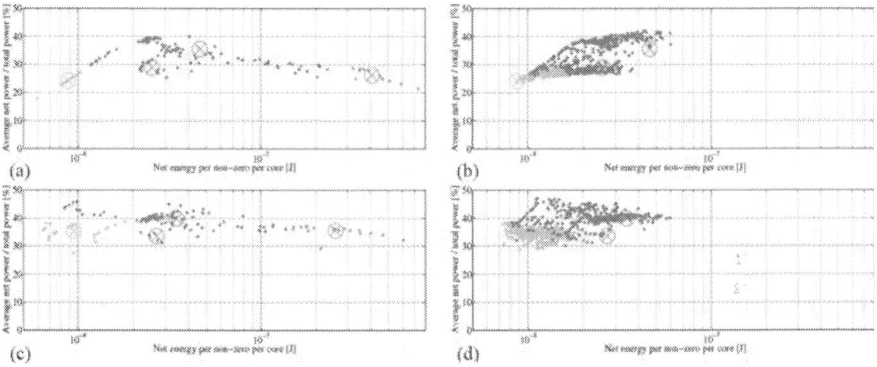

Figure 8. Matrix classification based on time, power, and energy measures on the IBM BG/Q. The position of the centroid of each is marked with a big crossed circle.

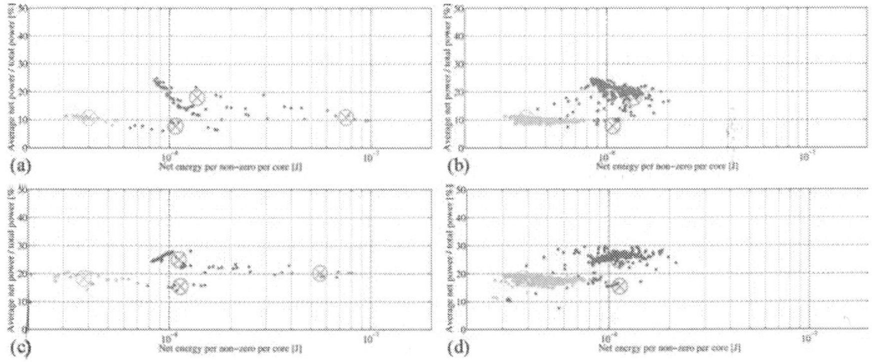

Figure 9. Matrix classification with respect to the triplet coordinated (*bs*, *bd*, *rd*) on the IBM P755.

We can also observe that from the point of view of performance: There is a vertical separation (both in power and energy) between the classes, depending on whether the problem data fits into the LLC or not. This is due to the additional energy required to move data to/from the DDR. The use of 4 threads per core increases the power but reduces the energy (mainly for the DDR cases). Energy increases linearly (in a log scale) with respect to time, while power generally decreases linearly. Inspecting **Figures 7a** and **8a**, we notice that, although real

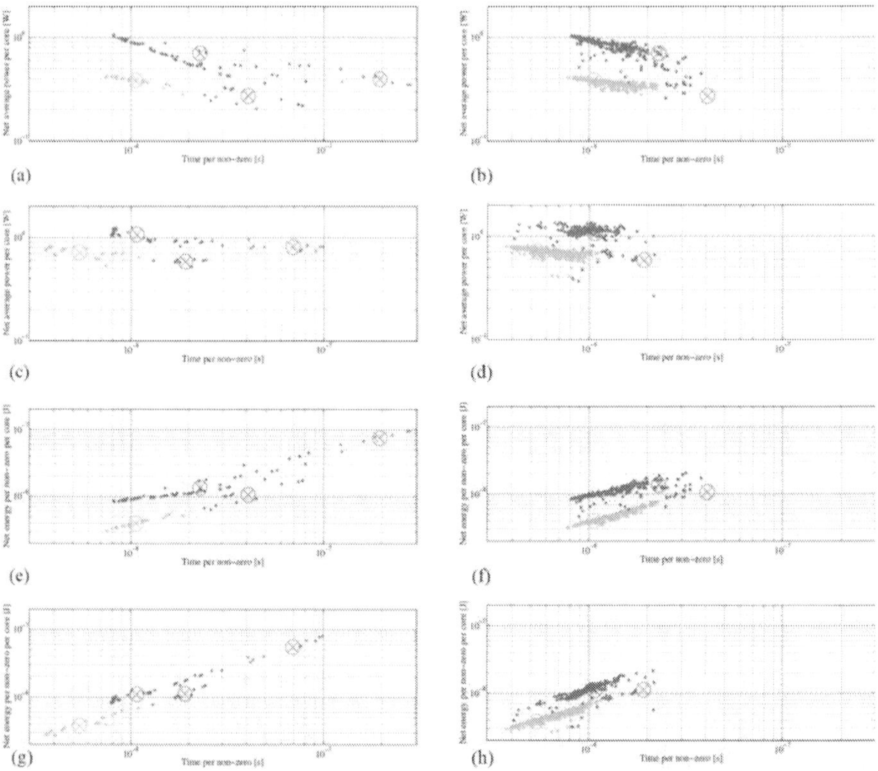

Figure 10. Percent of average net power consumption vs. net energy on the IBM BG/Q. Colors and centroids according to the matrix classification in **Figure 10**.

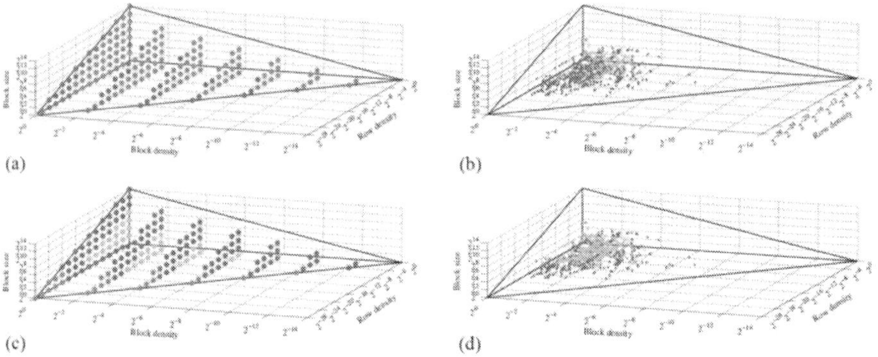

Figure 11. Percent of average net power consumption vs. net energy on the IBM P755. Colors and centroids according to the matrix classification in **Figure 12**.

matrices have a similar power vs. time behavior, energy performance is rather different. This hints to the fact that the FLOPS/W metric used in the Green500 with Linpack, is actually inappropriate to represent energy efficiency of general algorithms [21, 22]. Finally, **Figures 10** and **12** show a direct view over the power vs. energy consumption relation; here, the importance of the net power (especially on the P755) with respect to the overall power consumption is highlighted.

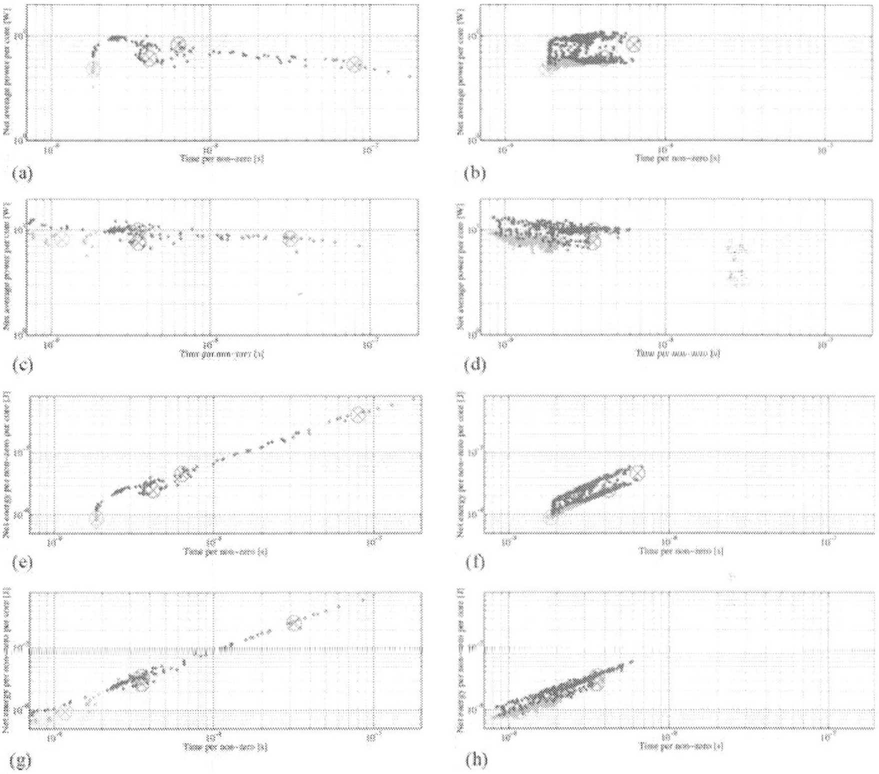

Figure 12. Matrix classification based on time, power, and energy measures on the IBM P755. The position of the centroid of each class is marked with a big crossed circle.

4.3. Validation

We evaluate our classification using the entire University of Florida Sparse Matrix Collection [23]. From this benchmark of real applications, we exclude complex and nonsquare matrices, as well as matrices with empty rows and the cases that do not fit in the target architectures due to DDR capacity restrictions. The resulting number of matrices available for validation consists of 1193 and 1202 samples for IBM BG/Q and IBM P755, respectively.

Class	Color	Memory	Time	Power	Energy
IBM BG/Q					
1	●	Cache	Low	Medium	Low
2	●	Cache	Medium	Low-medium	Medium
3	●	DDR	Low-medium	Medium-high	Medium-high
4	●	DDR	High	Medium	High
IBM P755					
1	●	Cache	Low	Low-medium	Low
2	●	Cache	Medium	Medium	Medium
3	●	DDR	Low-medium	Medium-high	Low-medium
4	●	DDR	High	Low-medium	High

Table 1. Qualitative behavior of each class of matrices.

For both architectures, we have measured time, power, and energy of the entire collection. This is displayed in the right-hand side plots of **Figures 10** and **12** (see also **Figures 7b** and **8b** for a 3-D representation), where colors (classes) have been assigned using the same *k*-means classification applied to the training sets (indeed, note that the centroids are the same in the left- and right-hand side images). The cross comparison of colors (i.e., classes) and values (i.e., time-power-energy measures) between the left and right graphs demonstrates that our classification strategy (based on coarse regular training sets) accurately captures a broad range of matrices from real applications for both architectures under investigation. In other words, the classification in **Table 1** offers a cheap, reliable, fast, and simple strategy to qualitatively determine time, power, and energy consumption for SpMV.

4.4. Conclusions

HPC supercomputers, and particularly cloud computing centers, can tremendously benefit from the existence of automatic tools to administer and optimize execution performance and cost of continuous and large stream of parallel tasks (from many different users). We have devised a systematic machine learning algorithm to classify and predict the performance costs of the SpMV kernel. The validity, accuracy, and robustness of our strategy have been demonstrated over a wide database of real matrices. More importantly, behind all the technical details, this approach actually represents a first concrete step towards the development of a global tool, which is able to characterize and capture the features of any sparse linear algebra operation (with a straight-forward extension to the dense case), thus covering a significant percentage of existing scientific computing kernels.

5. General-purpose multi-core servers: energy-aware runtimes for (sparse) linear algebra

Linear algebra operations and, in particular, sparse linear systems are a fundamental building block in many scientific and engineering applications. It is not surprising, therefore, that considerable effort has been spent by the scientific and high performance community towards the design and optimization of numerical methods for the solutions of these sparse problems, which can exert significant impact on the efficient solution of the applications that are built upon them.

The conjugate gradient (CG) method is among the most effective Krylov subspace solvers for sparse symmetric positive definite (SPD) linear systems [8, 9]. Furthermore, when the problem features only a few right-hand side vectors, the method has been also proven to be highly competitive for the solution of dense linear systems [21]. In this section, we illustrate the positive effects of integrating energy efficiency techniques into numerical software for linear algebra by considering the solution of sparse linear systems via ILUPACK (incomplete LU package) [24] as a workhorse. ILUPACK is a numerical software based on Krylov-based iterative methods that implements multilevel ILU preconditioners for general, symmetric indefinite, and Hermitian positive definite systems, in combination with inverse-based ILUs and Krylov subspace solvers. In other words, ILUPACK provides an implementation of the CG method (for sparse SPD linear systems) furnished with a very sophisticated preconditioner.

5.1. Energy-efficient processor technology

Current processors have adopted tools and technologies, originally designed for embedded systems and the mobile market in order to improve the energy efficiency. As of today, most processors adhere to the advanced configuration and power interface (ACPI) standard [25], which allows to configure the system state depending on the workload, and thus offers a tool to tune the power usage to the actual needs. Concretely, the CPU defines two types of energy-related states, which are reviewed next.

P-states. The ACPI standard defines a collection of operating voltage-frequency pairs for the processor, referred to as the *performance states* (P-states). The number of P-states and their granularity (processor vs. core) depends on the specific processor architecture. For example, for some architectures, the P-states can only be set for all cores in the socket (e.g., Intel Xeon E5504), but in others each core can be set to a distinct P-state (AMD 6128).

Table 2 shows the voltage-frequency pairs associated to the different states available in two recent multi-core processors. As a general rule, increasing the frequency of compute-bound operations reduces the execution time. For memory-bound operations, tampering with the frequency could be expected to have no effect on the execution time. However, in some processors (e.g., the AMD 6128), the memory bandwidth is connected with the processor frequency, and increasing/reducing the frequency can actually have an analogous effect in the execution time.

C-states. The ACPI standard also defines the *processor* or *CPU power states* (C-states). This energy-saving mechanism determines the conditions to turn off certain parts of the processor. State C0 corresponds to a processor that is in normal mode of operation. Compared with this, in all other states (C1, C1E, C2...), power is shut down for certain components such as the lower levels of the memory hierarchy. A deeper C-state will surely save power, but also increases the latency to transition back to the active C0 state. The programmer can only set the appropriate conditions in the application so that, when the processor is idle, the operating system promotes it to an energy-efficient C-state.

For the examples in **Table 2**, the AMD 6128 presents three C-states: C0, C1, and C1E; the Intel Xeon E5504 has four C-states: C0, C1, C3, and C6.

	Intel E5504 (4 cores)		AMD 6128 (8 cores)	
	Voltage (V)	Frequency (GHz)	Voltage (V)	Frequency (GHz)
P0	1.04	2.00	1.23	2.00
P1	1.01	1.87	1.17	1.50
P2	0.98	1.73	1.12	1.20
P3	0.95	1.60	1.09	1.00
P4	n.a.	n.a.	1.06	0.80

Table 2. P-states of two recent processors, from Intel and AMD.

5.2. Task-parallel runtimes for linear algebra operations and general applications

In recent years, a number of runtimes have been proposed to alleviate the burden of programming multi and many threaded platforms. Concretely, OmpSs, StarPU, Mentat, Kaapi, and Harmony, among others, have followed the approach pioneered by Cilk, offering implicit parallel programming models with dependence analysis. When applied to dense linear algebra (DLA) operations, SMPSs (a precursor of OmpSs), StarPU, Quark, and SuperMatrix have demonstrated the advantage of extracting task-parallelism using this "runtime approach". The idea underlying these runtimes to leverage the task-parallelism of a DLA operation consists in decomposing the computations into a collection of tasks connected via data dependencies. In a subsequent stage, the runtime issue these tasks for execution following an out-of-order schedule that maximizes concurrency, while taking into account the dependencies; for example, see Ref. [26] for details. These ideas have been proven useful and the new OpenMP standard has integrated some preliminary primitives in order to integrate task parallelism [27].

5.3. Parallelizing ILUPACK on multicore processors

A potential approach to tackle the solver underlying ILUPACK consists in exploiting task parallelism via a runtime, yielding a dynamic schedule of the work to the cores, with numeric properties similar to those of the sequential ILUPACK. This approach is similar to those employed in DLA, but considerably more difficult due to the irregular data structures that are involved in an iterative sparse linear system solver.

Figure 13 reports the algorithm for the preconditioned conjugate gradient (PCG) method. There, the computation of the preconditioner, M, is the initial step of the solver (O0). The iteration, after this first step, is composed of a sparse matrix-vector product (SpMV, O1), the application of the preconditioner (O5), and several vector operations (dot products, axpy-like updates, 2-norm; in O2–O4 and O6–O8). For simplicity, we regard both the computation of the preconditioner and its application as "black boxes". In practice, these are by far the most difficult operations.

$A \rightarrow M$	O0. Preconditioner computation
Initialize $x_0, r_0, z_0, d_0, \beta_0, \tau_0; k := 0$	
while $(\tau_k > \tau_{\max})$	Loop for iterative PCG solver
$w_k := Ad_k$	O1. SpMV
$\rho_k := \beta_k / d_k^T w_k$	O2. DOT product
$x_{k+1} := x_k + \rho_k d_k$	O3. AXPY
$r_{k+1} := r_k - \rho_k w_k$	O4. AXPY
$z_{k+1} := M^{-1} r_{k+1}$	O5. Apply preconditioner
$\beta_{k+1} := r_{k+1}^T z_{k+1}$	O6. DOT product
$d_{k+1} := z_{k+1} + (\beta_{k+1}/\beta_k) d_k$	O7. AXPY-like
$\tau_{k+1} := \parallel r_{k+1} \parallel_2$	O8. vector 2-norm
$k := k + 1$	
endwhile	

Figure 13. Algorithmic formulation of the preconditioned CG method.

The concurrency implicit in the iterative solver underlying ILUPACK, can be explicitly exposed by applying nested dissection to the adjacency graph associated with the sparse coefficient matrix of the linear system. Concretely, by recursively applying a divide-and-conquer strategy to this graph, we can obtain a hierarchy of subgraphs that reveals the concurrency of the operation. The parallelism enabled by this splitting is in practice captured as a (directed) task-dependency tree, with nodes representing tasks and arcs specifying the dependencies between pairs of them, as shown in **Figure 14**.

The multithreaded execution of the task tree implicit in ILUPACK, on a multicore processor, can be then left in the hands of a runtime which dynamically maps tasks to threads (cores), taking into consideration the dependencies. The runtime can aim to improve different criteria, including, e.g., balancing the workload distribution during the computation of the ILU preconditioner and the subsequent solution of the triangular linear systems involved in the PCG (see **Figure 13**). The runtime keeps track of the ready tasks (i.e., tasks with all their dependencies fulfilled), which initially contain those tasks corresponding to the independent subgraphs (leaves of the tree). The threads update this information as they complete the execution of tasks allocated to them. From the implementation perspective, this information is maintained in shared data structures, and the concurrent access is carefully controlled via lightweight synchronization system calls.

In addition to the concurrency intrinsic to the calculation and application of the preconditioner, the operations that appear in the iterative PCG solve define a partial order which enforces a strict execution order. In the actual implementation, one can eliminate the explicit barriers between these operations by instead relying on a runtime which can accommodate the nested parallelism exhibited by the task/subtask dependencies. In such scenario, the nested variant defines O1 + O2, O3 + O4, O5, O7, and O6 + O8 as five coarsegrain tasks, and offloads the complete detection and control of the dependencies to the runtime. In addition, these five macrooperations can be further divided into fine grain subtasks, and merge pairs of them as described above.

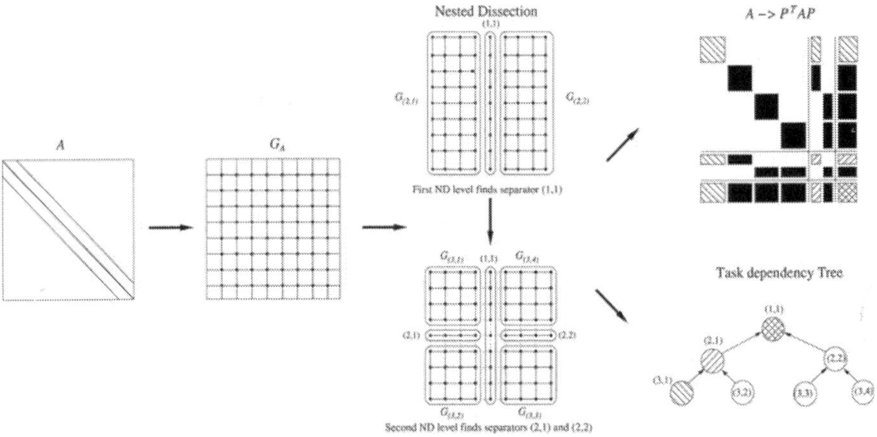

Figure 14. Nested dissection applied to the adjacency graph associated with a sparse matrix and corresponding task dependency tree.

For current NUMA architectures, the application records where (i.e., the socket) each task was executed on during the initial calculation of the preconditioner to achieve that tasks which operate on the same data that was generated/accessed during the preconditioner calculation and the PCG solve are mapped to the same socket [28]. This strategy ensures that, during the PCG iteration, a task is always executed on (any core of) the same socket that computed the corresponding task during the computation of the preconditioner.

For manycore processors, such as the Intel Xeon Phi, a critical aspect is how to bind the application threads to the hardware threads/cores of the systems, in order to attain a balanced distribution of the workload.

5.4. Optimizing energy consumption for linear algebra operations via the runtime

In general, optimizing energy consumption strongly requires the optimization of performance as a first step. This requisite particularly holds for DLA operations and especially the solution of sparse linear systems. For ILUPACK, though, it is possible to improve the energy efficiency of a runtime-based parallelization, with negligible impact on performance, by tuning the runtime itself to be aware of the ACPI C-states and P-states.

In a runtime-based parallelization, an idle runtime thread can rely on either polling or blocking policies upon encountering no task ready to be executed. For example, the prototype of SuperMatrix enforced a "busy-wait" for idle threads, till a new task is available. Unfortunately, this option impedes the transition of the corresponding core to a power-saving C-state because the thread is active (though doing useless work). From the positive point of view, this approach favors an immediate reaction of the thread to the creation of a new task. Alternatively, a power-aware version of the runtime can enforce an "idle-wait" (blocking) for idle threads.

The exploitation of the P-stages can also yield some potential energy savings if correctly adjusted from the runtime. Unfortunately, in some platforms, the use of the P-states is constrained by the operation of this mechanism at the socket level (i.e., it is not possible to change the P-state of a single core; instead, the change must be applied simultaneously for all cores of the socket). A second limiting factor for the P-states is the interplay between the frequency and memory bandwidth for some processor architectures, which implies that a reduction of the former negatively affects the latter as well. In consequence, exploiting this mechanism is far more delicate than doing so with the C-states. In particular, as energy consumption equals the product of time and power dissipation, the use of a lower P-state that reduces the frequency, and in principle, also the power draft, can result in a longer execution time that blurs the benefits from the point of view of energy consumption. In other words, reducing of power consumption is beneficial only if it does not increase the execution time that destroys the positive effects on energy consumption.

6. Energy-efficient techniques for multigrid methods on distributed memory platforms

Multigrid solvers belong to the most efficient numerical methods for solving symmetric positive definite linear systems. The computational complexity is $O(n)$ for sparse systems with n unknowns. To introduce the multigrid methodology, we revisit the definition of an iterative linear solver. Let $x^{(k)}$ be the approximation to the solution x resulting from the k-th iteration of the solver. The (unknown) error is denoted as $e^{(k)} = x - x^{(k)}$ and the residual vector is defined as $r^{(k)} = b - Ax^{(k)}$. Note that $e^{(k)}$ solves the error equation $Ae^{(k)} = r^{(k)}$ and $x^{(k)} = x \Leftrightarrow e^{(k)} = 0 \Leftrightarrow r^{(k)} = 0$. The abstract solution scheme stated in Algorithm 1 is known as iterative refinement. This scheme offers full flexibility for choosing a method to solve the error equation in step 4. Convergence of the approximation to the solution is guaranteed if the correction is sufficiently accurate since $r^{(k+1)} = r^{(k)} - Ac^{(k)}$. The idea of multigrid methods is to compute the error correction in step 4 of the iterative refinement scheme, not in the same space where the final solution is sought, but in a space of smaller dimension. The theoretical basis of multigrid methods is formulated in terms of "subspace correction" methods, see Refs. [29, 30]. Two prototypical variants of multigrid methods can be distinguished: the algebraic multigrid (AMG) variant is based on a purely algebraic problem formulation by means of a linear system of equations [31]. In contrast, the geometric multigrid (GMG) variant is based on the discretization of the underlying model equations on several grid refinement levels [32]. Both multigrid variants use a hierarchy of grid levels, either derived from the algebraic problem formulation or resulting from the discretization of the model equations on different computational grids. The process of constructing the hierarchy from the finest level is called "coarsening". The transition between two grid levels is defined by means of grid transfer operators, namely restriction and prolongation. Smoothing methods are used in order to make the unknown error from a fine grid representable on a coarser grid, where the correction shall be computed. The smoother is applied to the approximation on the fine grid and has the implicit effect of removing high frequency contributions from the error. The multigrid cycle is stated in Algorithm 2.

Algorithm 1 Iterative refinement

Set initial solution $x^{(0)}$, tolerance $\delta > 0$, iteration counter $k = 0$.
Compute initial residual $r^{(0)} = b - Ax^{(0)}$.
while $\| r^{(k)} \| > \delta$ **do**
Solve error equation $Ae^{(k)} = r^{(k)}$ approximately by means of a correction $c^{(k)} \approx A^{-1} r^{(k)}$.
Update solution $x^{(k+1)} = x^{(k)} + c^{(k)}$.
Compute residual $r^{(k+1)} = b - Ax^{(k+1)}$.
$k \leftarrow k + 1$.
end while

Algorithm 2 Cycle(A_h, x_h, b_h, γ)

1: **if** $h = H$ **then**
2: $x_h \leftarrow A_h^{-1} b_h$ (coarse grid solution)
3: **else**
4: $x_h \leftarrow S_h(A_h, x_h, b_h)$ (presmoothing)
5: $r_h \leftarrow b_h - A_h x_h$ (residual computation)
6: $b_{2h} \leftarrow R_{2h}^{h} r_h$ (restriction)
7: $x_{2h} \leftarrow 0$
8: **for** $k = 1, 2, \ldots, \gamma$ **do**
9: Cycle$(A_{2h}, x_{2h}, b_{2h}, \gamma)$ (recursion)
10: **end for**
11: $c_h \leftarrow I_{2h}^{h} x_{2h}$ (prolongation)
12: $x_h \leftarrow x_h + c_h$ (correction)
13: $x_h \leftarrow S_h(A_h, x_h, b_h)$ (postsmoothing)
14: **end if**

It is a recursive algorithm with its recursion basis on the coarsest grid level, where the error equation is solved with high accuracy in step 2. On all finer levels, only smoothers, residual computation, and grid transfer operators are used. The typical choices for the recursion parameter $\gamma = 1$ or $\gamma = 2$ lead to the V- or W-cycle, respectively, depicted in **Figure 15**. The solution is sought on the finest grid level where one expects the highest accuracy. The problem size in terms of number of variables is the largest on the finest grid level, and becomes subsequently smaller on the coarser levels. Standard smoothing methods include classical relaxation schemes like Jacobi, (symmetric) Gauss-Seidel or (symmetric) successive over-relaxation, and many other smoothing methods have been developed for specific fields of application. The smoother and grid transfer operator computations on fine grid levels usually only employ vector operations, sparse matrix-vector multiplications or element-wise operations in the grid. Only on the coarsest level, a direct or iterative method is used to solve the error correction equation with high accuracy. In the following paragraphs, we briefly introduce distributed memory platforms, the domain decomposition parallelization, which is often used on such platforms, and some implications on the numerical linear algebra routines.

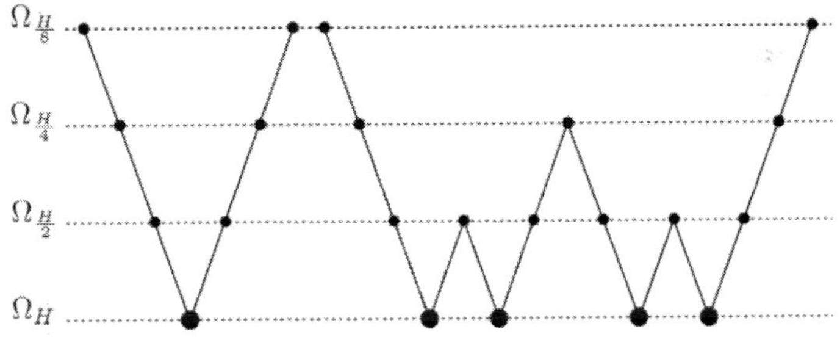

Figure 15. V-cycle (left) and W-cycle (right) across four grid levels. Small dots indicate the execution of the smoother, residual computation and grid transfer operators, while large dots indicate the solution of the coarse grid error equation.

6.1. Distributed memory platforms

Distributed memory means the separation of the available memory in distinct address spaces. This may be inherent to the computer platform, e.g., the computer units, or nodes, in interconnected clusters often have their own memory address space which is not accessible from other nodes. A distributed memory setup may also be induced by the coexistence of several host processes with individual private address spaces within the same computer. This is a fundamentally different situation compared to shared memory platforms, where all host processes or threads can access the same memory using a common address space. Note that there exist techniques which create a shared memory view of actually distributed memory platforms, but this is not discussed here. Parallelism on distributed memory platforms can be exploited by means of the "single instruction multiple data (SIMD)" or "multiple instruction multiple data (MIMD)" paradigm. It means that the problem data is split into pieces and distributed, while program replica (SIMD) or individual programs (MIMD) work on the data pieces in parallel. The processes which constitute the kind of parallel scientific application which we consider, may run on separate compute nodes in an interconnected cluster, possibly with several processes per node. Unless the application is "embarrassingly parallel", which denotes a situation without any dependency between the processes, a technique for making data from one process available to other processes is needed. The widely used standard technique in HPC is explicit data transfer between processes using the message passing interface (MPI) [33].

6.2. Domain decomposition parallelization and its implications on linear algebra routines

The distributed memory parallelization technique used for many numerical simulations of physical processes is based on domain decomposition. Using a number p of processes, the computational grid Ω_h is divided into a corresponding number of subdomains $S_q \subset \Omega_h$, $\bigcup_{q=1}^{p} S_q = \Omega_h$. In practice, graph partitioner tools can be used to achieve a balanced decomposition. **Figure 16** shows an example of a domain decomposition of a flow channel with an obstacle into eight subdomains.

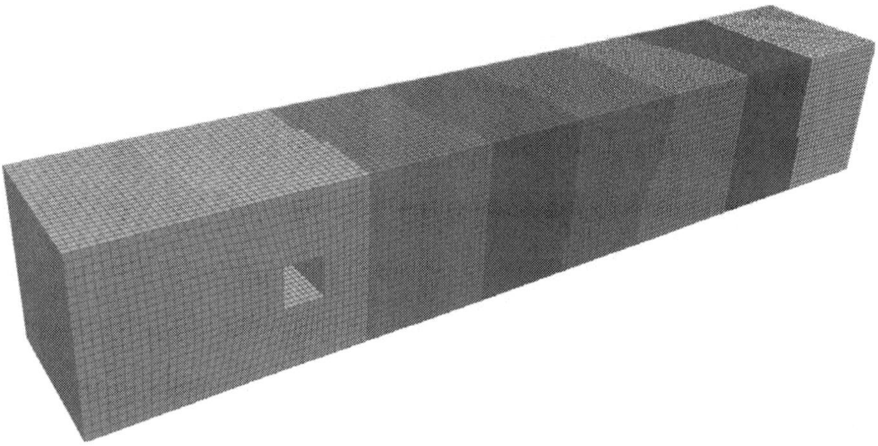

Figure 16. Domain decomposition of the computational grid of a flow channel with an obstacle into eight subdomains.

The domain decomposition implies a distribution of the discrete variables since these are associated to certain locations in the grid. Usually one uses a process-wise enumeration of the variables, so that the vectors and matrices of the discrete system are distributed in contiguous blocks among the processes. The usual technique to account for couplings of variables across subdomain boundaries is the addition of a "ghost layer" or "halo" of grid cells. The ghost layer replicates cells with coupling variables from the subdomains of other processes. The replicated variables are also called "ghost variables", and they are used in read-only mode for contributions of remote processes to the computations of the local process. Therefore, if the computation of a process relies on recent contributions from other processes, all affected ghost variables need to be updated by means of a data transfer. The communication pattern for a ghost update is determined by the neighborhood relation of the subdomains and the couplings of the variables. The local nature of the couplings, resulting from the discretization techniques described above, maintains the locality of the communication pattern of each individual process, i.e., it comprises only the subdomain neighbor processes. To illustrate the effect of the domain decomposition parallelization on the actual computations using distributed matrices and vectors, we introduce some dedicated notation. The local part of a vector v that comprises the variables associated with the subdomain of the process q is denoted as v_q^{loc}, and the ghost variables of process q form the ghost vector part v_q^{ghost}. Accordingly, the distribution of a matrix A yields a matrix block on the diagonal A_q^{diag}, whose entries represent couplings among the local variables of process q, and an off-diagonal block $A_q^{\mathrm{off\text{-}diag}}$, which comprises couplings of the local variables of process q with variables from other processes, i.e., with ghost variables. **Figure 17** shows a distributed matrix with diagonal and off-diagonal blocks, and a distributed vector with local parts. The ghost vector parts are not shown in this graphic.

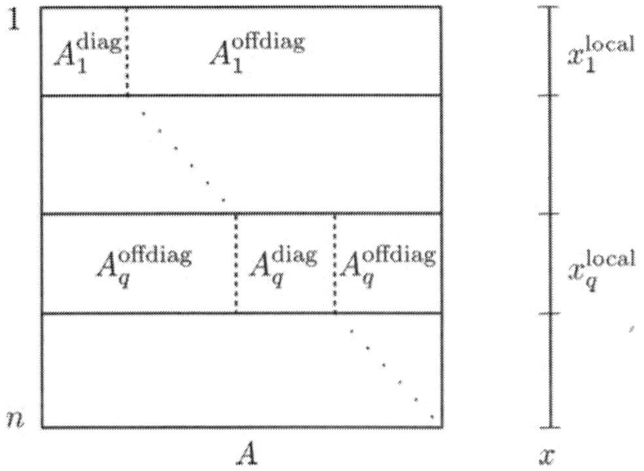

Figure 17. Distributed matrix with diagonal and off-diagonal blocks, and distributed vector local parts.

6.3. Parallel setup of some linear algebra routines

We can now formulate the distributed memory parallel version of some of the most important linear algebra routines, which are often used in numerical solvers. The scaling and addition of vectors can be done independently by each process for its local vector part, without using the ghost part:

$$\forall\ 1 \leq q \leq p : (\alpha v + w)_q^{\mathrm{loc}} = \alpha v_q^{\mathrm{loc}} + w_q^{\mathrm{loc}}, \tag{10}$$

where α is a scalar, and v and w are distributed vectors.

The matrix-vector multiplication uses the ghost vector part and therefore needs a ghost update:

$$\forall\ 1 \le q \le p : (Av)_q^{\text{loc}} = A_q^{\text{diag}} v_q^{\text{loc}} + A_q^{\text{off-diag}} v_q^{\text{ghost}}. \tag{11}$$

The ghost update needs to be completed before the off-diagonal matrix part is multiplied with the ghost vector part. However, the contribution from the diagonal matrix block multiplied with the local vector part is independent from other processes. This offers the opportunity to use a nonblocking communication mechanism for overlapping communication and computation, according to the following scheme:

1. Start nonblocking communication for the ghost vector update.
2. Compute $w_q^{\text{loc}} \leftarrow A_q^{\text{diag}} v_q^{\text{loc}}$.
3. Wait for completion of ghost vector update.
4. Compute $w_q^{\text{loc}} \leftarrow w_q^{\text{loc}} + A_q^{\text{off-diag}} v_q^{\text{ghost}}$.

In this scheme, the computation for step 2 potentially overlaps with the communication for the ghost update. This is advantageous since computation and communication may happen concurrently, thus accelerating the routine. However, it depends on the compute node and network hardware and system software to what degree the overlapping actually happens. If the node architecture is able to delegate the communication to the system, and meanwhile continue with the computation for the local contribution, a benefit for the overall performance can be expected. Ideally, the data transfer would happen entirely in the background, thus being completely hidden behind the computation overlap, without any communication overhead degrading the parallel efficiency. Of course, this can only be realized if the ghost update is completed before the computation for the local contribution finishes, so that there is actually no waiting necessary in step 3 of the scheme.

Without involving the ghost vector part, but nevertheless different from scaling and vector addition, and also from matrix-vector multiplication is the computation of scalar products and norms:

$$v \cdot w = \sum_{q=1}^{p} v_q^{\text{loc}} \cdot w_q^{\text{loc}}, \quad \|v\|_k = \left(\sum_{q=1}^{p} \|v_q^{\text{loc}}\|_k^k \right)^{1/k} \quad \text{for } 1 \le k < \infty, \tag{12}$$

$$\|v\|_\infty = \max_{1 \le q \le p} \|v_q^{\text{loc}}\|_\infty \tag{13}$$

The difference from scaling, vector addition, and matrix-vector multiplication is that the result has a global nature. Scalar product and norms require a global reduction operation such as global sum or global maximum. All processes must contribute to the result, and this result must be available on all processes. Therefore, scalar product and norm computations imply a global synchronization of all processes. Communication libraries such as MPI often provide routines with an optimized routing strategy, e.g., using tree-based routing algorithms of logarithmic complexity with respect to the number of processes, to implement global reduction operations.

6.4. Energy-efficient techniques for multigrid methods

Opportunities for making multigrid solvers more energy-efficient can be sought in all building blocks of the method including the smoother, grid transfer operators, and coarse grid solver. However, the individual parts must usually be addressed by individual measures for optimization due to their different roles in the multigrid algorithm.

As explained above, the purpose of the smoother is to remove high frequency contributions from the unknown error. To achieve this smoothing effect, the smoother does usually not need to yield an accurate approximation of the solution. Moreover, it does not always need to converge at all, as long as it has the desired smoothing properties. This gives space for the choice and optimization of smoothers. We depict one example in the following:

Traditional smoother choices include the classical Jacobi or Gauss-Seidel iteration and their damped variants. For a system matrix A with nonzero diagonal elements, the Jacobi iteration reads in component-wise form

$$x_i^{k+1} = x^k + \frac{1}{a_{ii}} \left[b_i - \sum_{j \neq i} a_{ij} x_j^k \right] \quad (i = 1, \ldots, n), \tag{14}$$

where k is the iteration index. The new iterate x^{k+1} relies on the previous iteration x^k, thus imposing a sequential order for the computation of the iterates. Moreover, in parallel setups, computing iterates usually involve contributions from other processes. This requires a synchronization of the processes after each iteration to make sure all needed values are updated from the last iteration.

For smoothers, it might, however, be acceptable to relax this strictly synchronized scheme by allowing to use also older or newer iterations in the computation. This leads to asynchronous iterations, which can benefit from massively parallel architectures like manycore devices or graphics processing units (GPUs). On the mathematical level, this is achieved by introducing a shift function, s, and an update function, u, in the algorithm:

$$x_i^{k+1} = \begin{cases} x_i^k + \frac{1}{a_{ii}} \left[b_i - \sum_{j \neq i} a_{ij} x_j^{k-s(j)} \right] & \text{if } i = u(k), \\ x_i^k & \text{if } i \neq u(k). \end{cases} \tag{15}$$

This abstract scheme can be adapted to fit the hardware at hand, e.g., by aggregating the components into blocks and mapping them to the cores of a multicore processor or to the thread blocks of a CUDA GPU. Thus, the adaption of the classic, synchronized relaxation scheme allows to efficiently exploit the parallelism of modern hardware, and in particular, it offers an opportunity to benefit from the superior FLOPS per Watt characteristics of GPUs.

In contrast, the residual and the grid transfer usually need to be computed accurately, because otherwise the overall convergence of the multigrid solver cannot be guaranteed. Nevertheless, performing residual and grid transfer computations on coprocessors such as manycore accelerators or GPUs might still prove beneficial, but care must be taken to ensure consistency in distributed systems. The coarse grid error correction solver is also expected to provide an accurate result. It nonetheless offers space for optimization, both in the choice of the method and its implementation. One usually employs Krylov subspace methods such as CG or GMRES methods, or direct methods such as LU or QR decompositions. The coarse grid solver can itself be subject to energy and performance optimization, and the overall multigrid solver would then inherit the benefits.

Another direction for energy and performance optimization, which is particularly relevant for distributed memory platforms, affects the parallel setup of the grid levels in the multigrid hierarchy. The problem sizes in the hierarchy often differ in several orders of magnitude from the largest problem size on the finest level to the smallest problem size on the coarsest level. A simple parallelization, where all grid levels are distributed to all available processors, may turn out to scale poorly for a large number of processors. This is due to the communication overhead becoming significant and diminishing the efficiency of the computations on coarse levels with small problem sizes. Instead, parallel setups, where coarser levels use only a subset of the available processors, can be beneficial. Balanced setups can maintain the overall performance of the parallelization, while reducing the communication, such that the overall efficiency is conserved. A fraction of the available processors can be temporarily deactivated, while the multigrid algorithm operates on coarse levels, and activated again to use the full computing power on finer levels. It is crucial to keep communication patterns local between neighboring subdomains, both within each grid level as well as between grid levels for the grid transfer. Well-configured parallel setups in the multigrid hierarchy can yield substantial energy savings by deactivating processes and reducing communication, while maintaining the overall performance.

Author details

Martin Wlotzka[1]*, Vincent Heuveline[2], Manuel F. Dolz[3], M. Reza Heidari[4], Thomas Ludwig[4], A. Cristiano I. Malossi[5] and Enrique S. Quintana-Orti[6]

*Address all correspondence to: martin.wlotzka@uni-heidelberg.de

1 Interdisciplinary Center for Scientific Computing (IWR), Heidelberg University, Germany

2 Computing Center, Heidelberg University, Germany

3 Computer Science Department, University Carlos III of Madrid, Spain

4 Department of Informatics, Universität Hamburg, Germany

5 Cognitive Computing and Computational Sciences Department, IBM Research, Zurich, Switzerland

6 Departamento de Ingenieria y Ciencia de Computadores, Universidad Jaime I, Castellon, Spain

References

[1] N.S. Trudinger D. Gilbarg. *Elliptic Partial Differential Equations of Second Order. Classics in Mathematics*. Springer Berlin Heidelberg, 2001.

[2] L.C. Evans. *Partial Differential Equations*, volume 19 of *Graduate Studies in Mathematics*. American Mathematical Society, 2 edition, 2010.

[3] R.J. LeVeque. *Finite Difference Methods for Ordinary and Partial Differential Equations*. Society for Industrial and Applied Mathematics, Classics in Applied Mathematics edition, 2007.

[4] J.L. Guermond A. Ern. *Theory and Practice of Finite Elements*. Springer New York, 2010.

[5] R. Herbin R. Eymard, T. Gallouete. *Finite Volume Methods. In: Handbook of Numerical Analysis, Vol. 7*. Elsevier. 2000.

[6] G. Wanner E. Hairer. *Solving Ordinary Differential Equations II: Stiff and Differential Algebraic Problems*. Springer Berlin Heidelberg, 2010.

[7] G. Wanner E. Hairer, S.P. Norsett. *Solving Ordinary Differential Equations I: Nonstiff Problems*. Springer Berlin Heidelberg, 2008.

[8] C.F. van Loan G.H. Golub. *Matrix Computations*. Johns Hopkins University Press Baltimore London, 4 edition, 2013.

[9] Y. Saad. *Iterative Methods for Sparse Linear Systems*. Society for Industrial and Applied Mathematics, 2 edition, 2003.

[10] Towards supercomputers: Eu project Exa2Green improves energy efficiency in high performance computing. *ICT-Energy Lett.*, (10), 2015.

[11] S. Catalan M.F. Dolz G. Fabregat R. Mayo E.S. Quintana-Orti S. Barrachina M. Barreda. An integrated framework for power-performance analysis of parallel scientific workloads. In *Proceedings of the 3rd International Conference on Smart Grids, Green Communications and IT Energy-Aware Technologies (ENERGY)*, pages 114–119, 2013.

[12] A.D. Malony S.S. Shende. The tau parallel performance system. *Int. J. High Perform. Comput. Appl.*, 20, 2006.

[13] M. Jurenz M. Lieber H. Brunst H. Mix W.E. Nagel M.S. Müller, A. Knüpfer. Developing scalable applications with vampir, vampirserver and vampirtrace. In *Proceedings of the Parallel Computing: Architectures, Algorithms and Applications Conference (PARCO)*, 2007.

[14] *Lapack project homepage*.

[15] R. Mayo E.S. Quintana-Orti R. Reyes M. Barreda, M.F. Dolz. Binding performance and power of dense linear algebra operations. In *Proceedings of the IEEE 10th International Symposium on Parallel and Distributed Processing with Applications*, pages 63–70, 2012.

[16] *The Top500 List,* June 2014.

[17] *The Green500 List,* June 2014.

[18] B.C. Catanzaro J.J. Gebis O. Husbands. K. Kreutzer D.A. Patterson W.L. Plishker J. Shalf S.W. Williams K.A. Yelick K. Asanovic R. Bodik. *The Landscape of Parallel Computing Research: A View from Berkeley.* Technical report, Electrical Engineering and Computer Sciences, University of California, Berkeley, Tech. Rep. UCB/EECS-2006-183, 2006.

[19] C. Bekas A. Curioni E.S. Quintana-Orti A.C.I. Malossi, Y. Ineichen. Performance and energy-aware characterization of the sparse matrix-vector multiplication on multithreaded architectures. In *Proceedings of the 43rd International Conference on Parallel Processing Workshops,* 2014.

[20] S. Lloyd. Least squares quantization in pcm. *IEEE Trans. Inf. Theory,* 28(2):129–137, 1982.

[21] A. Curioni C. Bekas. A new energy aware performance metric. *Comput. Sci. Res. Dev.,* 25 (3–4):187–195, 2010.

[22] P. Klavik, A.C.I. Malossi C. Bekas A. Curioni. Changing computing paradigms towards power efficiency. *Philos. Trans. Math. Phys. Eng. Sci.,* 372(2018), 2014.

[23] *The University of Florida Sparse Matrix Collection,* April 2014.

[24] *Ilupack,* July 2015.

[25] *Acpi,* July 2015.

[26] *Ompss,* July 2015.

[27] *The Openmp Api Specification for Parallel Programming,* November 2015.

[28] M. Barreda M. Bollhöfer E.S. Quintana-Orti J.I. Aliaga, R. Badia. Leveraging task-parallelism with ompss in ILUPACK's preconditioned CG method. In *Proceedings of the 26th International Symposium on Computer Architecture and High Performance Computing,* pages 262–269, 2014.

[29] J. Xu. Iterative methods by space decomposition and subspace correction. *SIAM Rev.,* 34 (4):581–613, 1992.

[30] H. Yserantant. Old and new convergence proofs for multigrid methods. *Acta Numer.,* 2:285–326, 1993.

[31] K. Stueben. A review of algebraic multigrid. *J. Comput. Appl. Math.,* 128:281–309, 2001.

[32] C.W. Oosterlee P. Wesseling. Geometric multigrid with applications to computational fluid dynamics. *J. Comput. Appl. Math.,* 128:311–334, 2001.

[33] *Message Passing Interface Forum. MPI: A Message-Passing Interface Standard,* Version 3.0. University of Tennessee, Knoxville, Tennessee, 2012.

Chapter 7

Energy-Efficient Communication in Wireless Networks

David Boyle, Roman Kolcun and Eric Yeatman

Additional information is available at the end of the chapter

http://dx.doi.org/10.5772/65986

Abstract

This chapter describes the evolution of, and state of the art in, energy-efficient techniques for wirelessly communicating networks of embedded computers, such as those found in wireless sensor network (WSN), Internet of Things (IoT) and cyberphysical systems (CPS) applications. Specifically, emphasis is placed on energy efficiency as critical to ensuring the feasibility of long lifetime, low-maintenance and increasingly autonomous monitoring and control scenarios. A comprehensive summary of link layer and routing protocols for a variety of traffic patterns is discussed, in addition to their combination and evaluation as full protocol stacks.

Keywords: Internet of Things, sensor networks, energy efficiency, communications protocols

1. Introduction

The promise of the 'fourth industrial revolution' relies on the development and integration of the so-called Internet of Things, cyberphysical systems and associated services and process improvements. The basis of the promise is the ability to instrument, connect, automate and remotely manage the majority of industrial systems and processes. The underlying assumption is that cheap, wirelessly communicating sensors and actuators can contribute to providing this capability, either autonomously or as part of a decision support system with a human in the loop.

In many cases, it is assumed that monitoring and control applications will require the use of devices that operate without persistent energy availability. Where no mains power is available, energy becomes a major constraint for applications expected to operate for increasingly long time periods. These periods are determined by the feasibility of maintenance of devices, with respect to both practicality and cost. Therefore, a major recent theme of research has been to attempt to achieve *energy neutrality*. This has been approached from many different angles, including novel ultra-low power receiver circuits (wake-up radio, WuR) [1, 2], energy harvesting and hybrid-storage (battery-supercapacitor) systems [3], compressive and predictive sensing [4–6], and traditionally, energy-efficient communications protocols [7].

Achieving energy neutral operation requires a comprehensive understanding of the application at design time and is seen as a 'holy grail' for networked embedded computing devices. However, applications tend to be characterised by heterogeneous performance requirements [8], deployment environments and criticalities. Therefore, there are few, if any, 'one size fits all' solutions. If one assumes that sensors and actuators are low cost with respect to energy in terms of sampling and information processing (which is not always true in the case of industrial applications [9]), communication is widely accepted to be the primary consumer of energy [10]. This energy consumption occurs when the radio transceiver is in an active state to send, receive and/or route packets. Assuming a worst-case scenario whereby the radio transceiver is always active (a state which tends to require tens of milliwatts of power [11]), the obvious way to reduce energy consumption is to place the device in the lowest power mode available for as long as possible—a technique known as duty cycling [12], which can be applied to the radio transceiver (radio duty cycling, RDC) and the other components of the device. This is equally true for recent system on chip (SoC) solutions that integrate transceiver and microcontroller circuitry on the same chip.

However, the device must also be able to participate in a network, which necessitates the implementation of a communications protocol stack (Section 1.1.1). This is a long-standing research area in the wireless sensor networks (WSN) community, which in the past 10–15 years has developed numerous energy-efficient communications mechanisms that operate at and across various layers. The remainder of this chapter is dedicated to exploring the evolution of energy-efficient communications primitives, explaining the inherent trade-offs in the design space and providing a comprehensive description of the state of the art. The emphasis is placed on link and routing layers.

1.1. Wireless communication

There are several well-known wireless communications technologies in popular use. These range from cellular, now on the verge of the 5th Generation (5G) [13], to WiFi (IEEE 802.11) and Bluetooth Low Energy (BLE) [14], among the most widespread[1]. Recent WiFi and BLE chip-sets are significantly lower power than their predecessors and are increasingly competitive for certain classes of energy-efficient 'IoT'-type applications, particularly in the consumer electronics market.

This chapter focuses on wireless communications protocols suitable for use with RF transceivers and SoCs typically considered ideal for low-power WSN-type applications, such as the (now obsolete) TI CC2420 [16], CC2538 (used in the recent OpenMote devices) [17], and the NXP JN5168. These are compliant with the IEEE 802.15.4 standard for low-rate wireless

[1]Full coverage of wireless communication fundamentals is available in [15].

personal area networks (LR-WPAN) [18], which is responsible for underlying much of the research in this area and has found its way into numerous industrial standards and specifications including ZigBee [19], WirelessHART [20] and ISA100.11a [21].

1.1.1. Communications protocol suites

A communications protocol suite specifies how data should be formatted (i.e. in packets), transmitted and received (channel access), and routed in a network[2]. Layers of abstraction are used to describe the various networking functions involved and vary somewhat depending upon the model of abstraction. For example, the OSI model specifies seven layers, whereas the TCP/IP model (Internet Protocol suite) specifies four. This is a particularly relevant issue in WSN design, as it is well known that efficiencies are best achieved by co-designing the layers [22], but they are often designed independently (e.g. RPL) [23] to enable interoperability. The most recent IETF stack proposed by the 6TiSCH working group, which is particularly suited to IoT, is shown in **Figure 1** and loosely maintains the TCP/IP model. In this case, IEEE 802.15.4 time synchronised channel hopping (TSCH) is used at the 'link layer' (which includes medium access control), 6top is an interface layer to the 6LoWPAN adaptation and compression layer (used to reduce the size of the IPv6 header such that it comfortably fits in the 127-bytes available in an IEEE 802.15.4 physical layer packet), and RPL (Section 3.2.5) at the Internet/routing layer. UDP is used at the transport layer to mitigate the overheads associated with TCP (i.e. end-to-end handshaking), and the Constrained Application Protocol (CoAP) [24] handles application layer functionality.

While the higher layers of the stack benefit from having an understanding of, and in many instances real-time information from, the lower layers, the most critical from an energy efficiency

```
+-----+-----+-----+------+-------+-----+
|     (COMI)      |(PANA)|6LoWPAN| RPL |
| CoAP  / DTLS    |      |  ND   |     |
+-----+-----+-----+------+-------+-----+
|       UDP       |       ICMP         |
+-----+-----+-----+------+-------+-----+-----
|               IPv6                       |
+------------------------------------------+
| 6LoWPAN adaptation and compression (HC)  |
+------------------------------------------+
|               6top                       |
+------------------------------------------+
|         IEEE802.15.4    TSCH             |
+------------------------------------------+
```

Figure 1. The 6TiSCH Protocol Stack [84].

[2]The practical implementation of a protocol suite is typically referred to as a 'stack'.

perspective are the physical layer (i.e. the radio transceiver itself) and the link layer, the latter of which is responsible for medium access control, and therefore, how long the transceiver remains in an active or sleep state. Energy-efficient medium access control (MAC) protocols are discussed in Section 2. However, as discussed in Section 2, the selection of a suitable MAC protocol is heavily dependent upon the application, specifically with regard to the statistical properties of the traffic generated in the network, the network topology and environmental factors. A well-designed application will take each of these factors into account at design time.

1.2. Hardware

The electrical characteristics of the hardware play a significant part in the overall energy efficiency of a networked device and the performance of a network. While a complete analysis of each component is beyond the scope of this chapter, it is worth highlighting where hardware mostly impacts the design of applications and implications for various layers in the protocol stack.

Firstly, selection of the radio itself is a key. From an application standpoint, basic quality of service requirements must be met, primarily bandwidth—otherwise the application is probably infeasible (without resorting to trickery in software, such as compressive or predictive sensing, and assuming this is satisfactory for an end-user). The worst-case scenario from an energy perspective is that the radio must be on 100% of the time. Therefore, once a wireless technology is selected (e.g. take IEEE 802.15.4), the designer sets about choosing a chip. The majority are reasonably similar in terms of their electrical characteristics, irrespective of the manufacturer. This makes life easier *but* it is worth considering the features of the chip selected. For example, on-board hardware acceleration of security functions can greatly reduce the overheads associated with implementation in software. Most contemporary RFICs compliant with IEEE 802.15.4 are in the tens of milliwatts range in active modes and drop to the microwatt range in low power modes, making them good candidates for long-term low-power applications. They require additional RF front-end design, including antenna and matching networks, plus an oscillator circuit to drive the clock. The latter requires the selection of inductors and capacitors, which are variable in their characteristics.

Important performance characteristics are often affected by environmental conditions such as temperature and humidity. In the case of a wireless node, these have effects internally (i.e. on each device) and externally (i.e. the effects on the wireless medium), which both impact on the performance of a network. In the case of the device, temperature effects and component selection have a significant effect on relative clock drift, which must be taken into account when tuning and learning protocol parameters like guard times and phase offsets, respectively (Section 2). Understanding clock drift is essential to tightly configure networking parameters, such as guard times to ensure accurate synchronisation, and has been studied in [25] where the authors investigate the effects of environmental temperature on clock drift and propose strategies to help designers choose optimal resynchronisation periods for given accuracies, and in [26] where the authors study the impact of oscillator drift on end-to-end latency over multiple hops using varying capacitor accuracies and show how to determine optimal parameters to minimise energy consumption in duty-cycled wireless sensor networks using

http://dx.doi.org/10.5772/65986

low power medium access control techniques. It is also worth noting that temperature influences battery performance, particularly as temperatures reduce, where capacity is degraded and voltage is known to reduce. This is a relatively understudied area in terms of IoT/WSN technologies, but is likely to be very important where devices are deployed outdoors in cold environments.

2. Energy-efficient medium access control

A large body of research exists concerning MAC protocols for WSNs. Notable examples of energy-efficient implementations include A-MAC [27], BoX-MAC [28], HuiMAC [29], SCP-MAC [10], ContikiMAC [30] and WiseMAC [31]. These MAC protocols are based on globally asynchronous, radio duty-cycled (RDC) approaches, where the objective is to minimise the active time of the RF transceiver. Typically, trade-offs are assumed to be inherent in the design of these protocols, and the Pareto-optimal solution is sought when considering energy efficiency and application level or performance requirements, such as throughput, reliability and latency. In [32], the authors consider low data-rate applications and attempt a tractable analytical approach to modelling latency and energy efficiency as functions of protocol parameters including duty cycle, slot duration and total slots, seeking to determine optimal settings for given workloads defined by application-level parameters. They find that WiseMAC best balances energy efficiency and latency based on the scenarios considered, and attribute minimising protocol overhead through local synchronisation (also referred to as *phase lock* optimisation in the literature, and exploited in several subsequent proposed MACs [30, 33]) and random channel access as key to achieving this balance.

These protocols, however, are just the tip of the iceberg (and are heavily oriented towards aggressively duty-cycled, energy-efficient scenarios for low data-rate applications). In [34], Bachir et al. present a comprehensive taxonomy of MAC protocols according to the various techniques being used, classifying them according to the challenge they address. They describe the importance of understanding the statistical properties of the network traffic when selecting and tuning a MAC protocol, which is argued to be a more useful approach for the application developer. The authors classify the traditional 'MAC families' as *Reservation-Based Protocols* and *Contention-Based Protocols* at the highest level, where the former is synonymous with scheduled approaches such as Time Division Multiple Access (TDMA), and the latter with simpler, popular approaches such as ALOHA and Carrier Sense Multiple Access (CSMA). They proceed to explore the functions relative to: high—*scheduled* protocols for high-rate applications such as multimedia; medium—protocols with *common active periods* for medium rate applications such as those found in many industrial applications; and low—*preamble sampling* for low-rate applications with rare event-reporting, such as long-term monitoring or metering, and finally consider *hybrid* protocols for time-varying application-level functionality. **Table 1** borrows the structure and updates the taxonomy of that presented in [34]. Specifically, A-MAC, Reins-MAC and IEEE 802.15.4e, which leverages TSMP and is closely linked with ongoing standardisation initiatives, are notable protocols published since Bachir et al. presented their study.

Functionalities	Protocols
Scheduled	TSMP, IEEE 802.15.4(e) [35], Arisha, PEDAMACS, BitMAC, G-MAC, SMACS, TRAMA, FLAMA, Î¼MAC, EMACs, PMAC, PACT, BMA, MMAC, FlexiMAC, PMAC, O-MAC, PicoRadio, Wavenis, f-MAC, Multichannel LMAC, MMSN, Y-MAC, Practical Multichannel MAC, LMAC, AI-LMAC, SS-TDMA, RMAC, Reins-MAC [36]
Common active period	SMAC, TMAC, E2MAC, SWMAC, Adaptive Listening, nanoMAC, DSMAC, FPA, DMAC, Q-MAC, MSMAC, GSA, RL-MAC, U-MAC, RMAC, E2RMAC
Preamble sampling	Preamble-Sampling ALOHA, Preamble-Sampling CSMA, Cycled Receiver, LPL, Channel Polling, BMAC, EA-ALPL, CSMA-MPS, TICER, WOR, X-MAC, MH-MAC, DPS-MAC, CMAC, GeRAF, 1-hopMAC, RICER, WiseMAC, RATE EST, SP, SyncWUF, STEM, MFP, 1-hopMAC, SpeckMAC-D, MX-MAC, IX-MAC [33], ContikiMAC [30], LPP [37], RI-MAC [38], A-MAC [27], Flip-MAC [39]
Hybrid	IEEE 802.15.4, ZMAC, Funneling MAC, MH-MAC, SCP, Crankshaft

Table 1. Summary of MAC protocols by functionality, updates [34].

Another interesting development relates to the exploitation of constructive interference, which is contrary to CSMA mechanisms and seeks to benefit from concurrent transmissions. Flip-MAC [39] and Glossy (not strictly a MAC protocol) [40], for example, make good use of this with regard to packet acknowledgements and efficient network flooding, respectively.

2.1. Standards and evolution

The IEEE 802.15.4 standard is one of the most important standards in WSN/IoT. First published in 2003, it specified the physical and medium access control mechanisms for low-rate wireless personal area networks and became part of the industrial standards and specifications listed in Section 1. The medium access control layer specified in the standard is effectively a hybrid construct that allows the use of slotted or un-slotted CSMA-CA with optional guaranteed timeslots and packet delivery. It was designed to accommodate a range of topologies, including star and peer-to-peer, specifying device classes that allowed for reduced protocol complexity for 'reduced' function devices (RFD) opposed to 'full' function device (FFD) capable of communication with any device in the network. While this allows for relatively low duty cycles to be achieved, there is an inherent trade-off between energy saving and latency and bandwidth. This was studied immediately, and simulations were used to illustrate the trade-offs related to using the various modes, for example, in [41].

De facto standards simultaneously emerged in the research community where a number of competing objectives made it difficult to build practical implementations for the hardware available at the time. Factors such as ultra-low power operation, collision avoidance, efficient channel utilisation, robustness to time-varying channel and networking conditions, scalability, and efficiency in terms of implementation and memory requirements needed to be addressed by design. B-MAC (short for Berkeley) was one of the earliest protocols built into an open source 'operating system' for WSNs, namely TinyOS [42]. B-MAC (2004) is a carrier sense protocol that improved upon the performance of S-MAC (one of the first to use RTS/CTS in sensor networking) under certain conditions, making use of clear channel assessments, acknowledgements, back-offs and low power listening (LPL). This was later improved upon by merging features with X-MAC (a link-layer only approach, 2006), resulting in the BoX-MAC (2008) protocols which eventually shipped with TinyOS [28]. BoX-MAC demonstrates cross-layer design and how to achieve significant energy efficiencies and throughput improvements for reasonable workloads.

Within the Contiki community, a number of the same protocols were implemented and extended into standard link layers and radio duty cycling mechanisms in a cross layer fashion. Features from BoX-MAC and X-MAC, such as periodic wake-ups and strobes, and WiseMAC, specifically the phase lock optimisation, were integrated to form the ContikiMAC protocol, which has shipped as a *de facto* standard link layer since around 2011.

TSMP is one of the most notable TDMA MAC layers developed in the last decade [43]. It uses synchronisation between devices to communicate in scheduled timeslots (allowing low-energy radio management) and operates reliably in noisy environments by using channel hopping to avoid interference, with different time-slotted packets sent on different channels depending on the time of the transmission. Therefore, the approach is suitable for applications that require relatively low-power and high-reliability performance, characteristic of industrial automation scenarios, and has thus been included in the WirelessHART and ISA standards (Section 1.1). More recently, it has become fundamental to the development of the IEEE 802.15.4e amendment to the standard, which is actively being included as a core technology in the IETF standardisation effort (e.g. **Figure 1**).

2.2. Design trade-offs

It is important to understand the trade-offs inherent in the link layer design choice. From an energy efficiency perspective, selection of a low-power transceiver is critical, but selection of the appropriate link layer will depend on the application's performance requirements. All of the aforementioned protocols are *relatively* energy efficient. Therefore, depending on the reliability, throughput and latency required, certain protocols perform better than others. For critical applications, it is usual that energy efficiency is not prioritised as highly as reliability and low latency. Therefore, in a high-criticality application, TSMP would be preferable to an asynchronous protocol such as ContikiMAC.

Table 1 shows that the area of MAC design for wireless sensor network applications has been comprehensively researched. As a result, many recently proposed protocols integrate and build upon ideas previously presented. The parameters of importance are now reasonably well known. CSMA-CA protocols with arguably the best performance typically operate using request-to-send/clear-to-send (RTS/CTS) signalling, where networked devices periodically listen to the channel to determine whether any neighbours want to send packets, or alternatively begin to strobe RTS packets and wait for a CTS message from a listening neighbour if the device wants to transmit. The choice of what to send in the RTS packet is an interesting one, with many designers proposing to effectively send an entire data packet as the RTS strobe, such as in ContikiMAC, whereupon the receiving node sends an ACK having received the data payload. However, this can be suboptimal considering variable length data packets. It is argued in [33] that the use of a fixed-length RTS packet can bring efficiencies when considering its relationship to strict timing parameters. Addressing information (to enable unicast, broadcast and/or multicast communication) is a key inclusion for such a packet.

Timing parameters for such protocols are bounded by performance characteristics of the RFIC (such as turnaround times between TX and RX modes) and the theoretical minimum times required to transmit, receive and act upon packets. Perhaps the most critical parameters are the *receive check* (RC) and *receive check intervals* (RCI). It is also worth noting that receive checks can be performed at both the physical (L1) and link layers (L2), which enforces a trade-off between absolute energy efficiency and reliability. Where L1 only methods are used, the RC operation is cheaper, but net energy efficiency may be worse if devices that are not being addressed stay in RX mode and parse packets not specifically intended for them.

The RC is where a node samples the wireless medium for energy to determine whether or not there is an incoming packet. An optimal duration for the RC is tightly coupled with the time taken to listen for a CTS strobe during a wake-up stream. Minimising the RC is a key to optimising energy efficiency, but this period is also closely related to reliability performance. Theoretical minimum values for length of the RC and the CTS listen period are calculated using the RXTX turnaround times for the transceiver, its bit rate and processing time.

The RCI is the interval between performing RCs, typically measured in Hz. It is possible to determine an optimal RCI with regard to energy efficiency, but this is detrimental to latency (where a latency is fundamentally lower bounded by the RCI and the number of hops over which a packet must travel to reach its destination). RCIs are often lengthened to reduce their impact on the quiescent power consumption of a device. Where large RCIs are used, for example, >0.5 Hz, it becomes more important to implement phase or offset learning (resulting from positive or negative relative clock drift) to ensure that energy consumption is minimised when beginning the next RTS/CTS wake-up stream.

Transmitting nodes may or may not, depending on the application, have an enforced periodic data reporting interval. This is often referred to as the inter-packet interval (IPI) in the literature and broadly reflects the frequency with which sensor data are transmitted from each device in the network. It is worth remembering that the IPI is not necessarily the same as the sampling rate of the sensor, where in many cases an aggregate or average values may be returned periodically (i.e. at the IPI rate of the node).

A factor less studied in the literature is the implementation of the CSMA algorithm itself. In [33], the authors demonstrate that a quantised approach to the implementation of back-offs performs better with regard to energy efficiency and reliability than randomised approaches.

So called receiver-initiated protocols have more recently been studied, whereby when a node wants to transmit a packet it waits for a neighbour (i.e. a potential receiver) to send a probe. Upon hearing the probe, the sending node transmits its packet. Examples of this approach are Low-Power Probing (LPP) as introduced in Koala [37], RI-MAC [38] and A-MAC [27].

The physical and link layers determine the lower bound for energy efficiency with respect to point-to-point communication. When a network extends beyond a one-hop (or star) topology, devices in the network must implement some routing functionality (usually referred to as

Layer 3 or the NWK layer), to ensure that packets arrive at the intended destination, which is explored in the next section.

3. Energy-efficient networking, data collection and dissemination

Routing algorithms are essential for every WSN as they define how packets flow within the network. Algorithms differ from each other based on their capabilities, effectiveness, amount of memory required to store the routing state, agility and energy awareness.

Routing protocols for WSNs can be categorised into three main groups depending on the functionality they provide: (*i*) protocols that route data only towards one or more predefined *base-station* or *sink* nodes, referred to as *data collection* protocols or *many-to-one* patterns of communication, (*ii*) protocols that allow *peer-to-peer* or *any-to-any* patterns of communication between nodes in the network and (*iii*) protocols allowing one node to *disseminate* a message to all or a subset of the nodes in the network, also referred to a *one-to-many* pattern of communication. Protocols which belong to the first group are usually based on building *routing trees* which are rooted at one (or more) sink node(s). Every node in the network forwards all data towards a sink via a selected parent. Protocols from the second group support peer-to-peer communication, that is, any node in the network can send a message to any other node. These protocols either exploit some kind of routing table, learned beforehand, to forward data towards the destination or they first need to find the destination node and establish a connection. Protocols from the last group are usually based on gossiping or flooding the whole network.

3.1. Data collection

Typically, WSNs have been deployed to collect data about certain phenomena in predefined geographic areas (sensing field). Nodes in such a network sample a sensor at a predefined rate and report the sensed data towards a sink node. The network may contain several base-stations, in which case a node typically forwards data towards the closest base-station only.

Two of the most common routing protocols are *Collection Tree Protocol* (CTP) [44] and *Routing Protocol for Low Power and Lossy Network* (RPL)[3] [23]. CTP is a standard library of the TinyOS [42] operating system, while the Contiki [46] operating system comes with two standard alternatives: (*i*) Collect, which is part of the basic Rime network stack [47] and is an alternative implementation of CTP for Contiki OS and (*ii*) RPL, which is part of the more complex IPv6 stack. There is also an RPL implementation for TinyOS called TinyRPL. These protocols are based on creating a directed acyclic graph rooted at the base-station. Each node in the network forwards all data towards the root.

A primary challenge in the design of these protocols is defining the metric by which a node chooses its parent, via which the node will forward all packets. Early adopters of this approach

[3]RPL is in fact designed to support any-to-any communication; however, very little work has been done in this regard, with early evaluations suggesting that it is not ready for actuation-type messages [45].

used hop count to the base-station as a metric [48]. Later, CTP used a new metric: Expected Transmission Count (ETX). ETX uses the number of transmissions required to deliver the packet to the destination without error. ETX depends on the quality of the link. Many protocols vary by how the quality of the link is measured and computed (e.g. based on RSSI, or a statistical measure of the number of packets lost, or a function of both). It has been shown that using this metric decreases the network traffic and leads to lower energy consumption.

Energy efficiency can be described as the ratio between the total number of packets received at the destination node (i.e. the base-station, in the case of collection protocols) and the total energy spent by the network to deliver these packets. Due to the overhead of IPv6 protocol, researchers were concerned about the energy efficiency of the RPL protocol. However, it was shown that packet delivery ratio of CTP and RPL is very similar, while the overall energy consumption is only 3% higher for the latter [49]. Similar results were observed in a study focused on interoperability of RPL implementation for Contiki and TinyOS operating systems [50].

A disadvantage of building rigid routing trees is that the nodes along the path towards the base-station have to transfer more data than other nodes, and hence their batteries deplete faster. This especially applies to the nodes closer to the base-station. To tackle this problem, Lindsey et al. proposed *Disjoint Paths* and *Braided Paths* algorithms [51]. Disjoint Paths algorithm constructs a small number of alternative paths from each sensor to the base-station. These paths are sensor-disjoint, that is, paths have no intermediate nodes in common. Braided Paths algorithm creates partially disjoint paths from the primary path, that is, for each node on the path an alternative path is created which does not contain given nodes. Therefore, in the case of a node failure, an alternative path can be used without the need to first find the path.

An alternative approach to creating rigid routing trees is the back-pressure protocol (BCP) [52] presented by Moeller et al. In networks with BCP, the routing decision depends on the size of the packet queue and the packet rate between two nodes. Each node maintains a queue of packets, where a base-station has a queue of zero length. A node forwards a packet to a neighbour only if the neighbour's queue is shorter than the queue of the sending node. The received packet is put on the top of the queue and in the next iteration forwarded to a node with a shorter queue. This can lead to more evenly spread network traffic while exploiting various routes towards the base-station.

An older approach which tries to eliminate rigid routing trees is *hierarchical routing*. This approach breaks the network into clusters, each of which has a *cluster-head*. Nodes send data to these cluster-heads which are then responsible for delivering data to the base-station.

Heinzelman et al. presented *Low-Energy Adaptive Clustering Hierarchy* (*LEACH*) [53], a popular clustering algorithm whose goal is to reduce energy consumption of nodes in a WSN. The operation of the algorithm is split into two phases: (*i*) the *setup phase* during which cluster-heads are elected and nodes choose which cluster they will be part of and (*ii*) the *steady phase* during which sensor nodes transfer data to the cluster-heads which aggregate received data and forward them to the base-station. Cluster-head election is distributed, and nodes do not require any global knowledge of the network. A disadvantage of this algorithm is that the communication between nodes and cluster-heads, as well as the communication

http://dx.doi.org/10.5772/65986

between the cluster-heads and the base-station is single-hop only, which limits the size of the network. Additionally, cluster-head selection does not take into account the residual energy of the node. Being a cluster-head is very energy demanding, as the cluster-head has to be 100% duty-cycled in order to receive messages from all nodes. If a node with small residual energy is elected, it may deplete its battery before it collects all data and forwards them to the base-station.

One of the extensions of LEACH algorithm presented by Lindsey and Raghavendra, *Power-Efficient Gathering in Sensor Information Systems (PEGASIS)* [54], creates chains of sensor nodes where each node aggregates data received from the previous node with its local data. In each iteration, a random node from the chain is chosen to forward aggregated data to the base-station. A disadvantage of PEGASIS is that it assumes that each node has global knowledge of the network layout, particularly the position of the nodes.

Younis and Fahmy presented another LEACH extension called *Hybrid, Energy-Efficient Distributed Clustering (HEED)* [55]. Unlike LEACH, it operates in multi-hop networks and for cluster selection it uses both the residual energy of the node and the node degree or density. Thus, the algorithm may better balance energy consumption amongst the nodes, hence prolonging the lifetime of the network.

Each of these hierarchical routing protocols scored three out of four points for energy efficiency in a large survey on routing protocols [56], meaning that these protocols achieved average packet delivery rates while choosing routes based on the residual energy, thereby prolonging the lifetime of the network.

3.2. Peer-to-peer routing

For networks deployed for the purpose of monitoring and continuous collection of data, peer-to-peer (P2P) communication is often not necessary. Each node only needs to know how to deliver data to one of the base-stations. However, as WSNs become more common and serve a wider range of purposes, communication among nodes in the network will become more important. WSNs are not only used to collect data but also to react to the environment and control it via *actuators*, that is, components of emerging cyberphysical systems. Nodes in the network need to send messages directly to other nodes, while lowering the overall traffic. With actuation networks, this requirement becomes even more important to send an actuation message directly to the actuator. For that purpose, routing protocols which allow P2P communication were developed.

Such P2P protocols may be categorised into five groups, depending on how they locate and forward messages to the communicate with a peer: (*i*) geographic routing, (*ii*) routing based on trees, (*iii*) hierarchical routing, (*iv*) ad hoc shortest path routing and (*v*) routing based on routing tables.

3.2.1. Geographic routing

In geographic routing, each node is not addressed by its ID or IP address but by its geographic location. The routing decision is then based on the position of the node making the forwarding decision, the position of the destination node and the position of the neighbours of the forwarding node. The neighbour which is closest to the destination node is chosen as the next-hop. As geographic routing heavily relies on the exact geographic position of the nodes, specialised hardware is required (e.g. GPS) and/or a localisation algorithm must be used. However, specialised hardware increases the price of the node, increases the energy requirements and is sometimes not very precise. Similarly, using localisation algorithms tends to lead to additional network traffic and may also be imprecise [57–59].

Among others, Karp and Kung proposed *Greedy Perimeter Stateless Routing (GPSR)* [60], Kuhn et al. proposed *Geometric Ad hoc Routing* [61], and Yu et al. proposed *Geographic and Energy Aware Routing (GEAR)* [62], all of which implement greedy geographic routing. Even though this type of greedy routing works well under ideal conditions, it fails when a *routing void* is encountered. The routing void is a situation when there is no neighbour which is closer to the destination node. In practice, this situation is common when there is either an error in the localisation algorithm or a physical obstacle prevents radio communication between nearby nodes. Additionally, there is currently no practical way of porting geographic routing to the three-dimensional space [63], for example, in a network deployed in a building or a tower.

According to the survey on routing protocols in [56], GEAR outperforms GPSR in terms of the packet delivery; however, it scores only two points out of four for energy efficiency.

3.2.2. Routing trees

In networks where P2P communication is based on *routing trees,* the nodes are organised in one or several trees where each node stores its parent's ID only. The root of a tree is represented by a more powerful node storing connectivity information of the whole network. The packet is first routed to the root of a tree, where the central router computes the shortest path to the destination. The packet is then routed downwards towards the destination via this shortest path. The advantage of this approach is minimal memory requirements on the nodes and simplicity of the routing algorithm. The disadvantage is potentially high *routing stretch,* that is, the ratio between the length of the found path and the optimal one, and the requirement that the central router is aware of the whole network topology. Additionally, top-level nodes may become overloaded by the network traffic, especially in large networks.

To tackle some of the disadvantages mentioned above, several improvements to the routing trees have been introduced. The key improvements are based on storing meta-data on the nodes within the network. Dedicated nodes store meta-data about all nodes in a sub-tree rooted in given node. Then, a node can decide to route a packet down a tree without forwarding it to the root node. In RPL, the base-station holds a routing table for the whole network. However, any node in the network, provided it has enough memory, can store a routing table for a sub-tree rooted in given node. These nodes are referred to as *routing nodes*. If a routing node receives a packet, it first checks its local routing table to see whether it contains a record for the destination. If so, the packet is forwarded directly to the destination. Otherwise, it is forwarded towards the root.

Duquennoy et al. presented an opportunistic version of RPL called *Opportunistic RPL (ORPL)* [64]. Here, each node uses a Bloom filter [65] to store all node IDs from the sub-tree rooted in a given node. Each node then uses the summary to decide whether the packet should be forwarded up towards the root of the tree or down the sub-tree rooted in a given node. When the packet travels up the tree, it does not necessary follow the spanning tree. Any node which is closer to the root can opportunistically forward the packet. These two improvements can significantly lower the routing stretch. However, the possibility of a false positive in Bloom filters is the main disadvantage of this approach. In this case, a special algorithm is required which can recover from the situation when a packet is routed down the tree based on a false positive. The packet has to be sent to the root of the tree which can then find the correct path to the destination. This leads to unnecessary traffic and large routing stretch.

Mihaylov et al. use a similar approach in their *Innet* algorithm [66], but, to reduce the routing stretch, they build up to three routing trees, each rooted in a different part of the network. Each node stores a summary for each sub-tree rooted in given node. The search for the destination node is performed in all routing trees in parallel. Furthermore, to avoid the problem of false positives in Bloom filters, the packet is always also forwarded up, until it reaches the root of a tree. The packet stores the path it takes until it reaches the destination node. The destination node replies to the source node by reversing this path. As the reply packet travels back to the source node, each node uses several techniques to find a shortcut between the communicating nodes. Innet was designed to support long-term communication, that is, the communicating peers exchange messages for a longer period of time. Therefore, the main goal of the algorithm is to minimise the routing stretch. The higher cost of the search phase is outbalanced by savings that could be achieved during the long-term communication amongst the nodes.

3.2.3. Hierarchical routing

In *hierarchical routing*, each node is a part of multi-level hierarchically organised clusters [63,67]. At the lowest level 0, each node is a member of its own singleton cluster. Then, a neighbourhood of level 0 clusters is organised into level 1 cluster, which in turn are grouped into level 2 cluster, until all nodes are member of one (or very few) big cluster(s). At each level, a node is a member of exactly one cluster.

At the centre of each cluster is a *cluster-head*. At each level i the cluster-head is advertised R_i hops away. The R depends exponentially on the level i. A node can be a member of a level i cluster if it is at most r_i hops away from the cluster-head, where $r_i \leq R_{i-1}$. In practice, usually $R_i = 2^i$ and $r_i = \lfloor R_i / 2 \rfloor$. Each node is addressable by concatenating the cluster-head ID at each level (e.g. $X.Y.Z$, where node's ID is X, Y is a level 1 cluster-head, and Z is a level 2 cluster-head).

The *routing table*, stored at each node, consists of entries for each cluster-head the node received an advertisement from. Remember that each cluster-head is at most R_i hops away at every level. Because the routing table is stored at every node, each node in a network acts as a router. When a node receives a packet, it tries to find the record in the routing table for the cluster-head from the lowest level. For example, if the packet's destination is $X.Y.Z$, the node first tries to locate a record for X. If it is not found, it tries the same for Y, and Z respectively. Because Z is the top-level cluster, every node in the network will know a route to it. The packet is routed towards the first found record. Because $r_i \leq R_{i-1}$ it is guaranteed that as the packet is routed towards the level i cluster-head, there will be a node on the path which knows the route towards the level $i-1$ cluster-head. Therefore, the packet will eventually reach the destination node. In practice, hierarchical routing achieves relatively small average routing stretch of 25% [63]; however, additional network traffic is required to keep the routing tables updated.

3.2.4. Ad hoc routing

Unlike other routing algorithms, *ad hoc routing* does not require any global preparation phase during which the network is prepared for P2P communication. However, when a node needs to communicate with a peer, a path between the nodes must be established first. This is usually done by flooding the network with a request [68–70]. The request contains the source node ID, the destination node ID and a path taken by the request so far. Each node, unless the node is the destination that receives the request, adds itself to the path and re-broadcasts the request. Once the destination node receives the request, it replies to the source node by reversing the path of relay nodes. The algorithm leads to discovery of the shortest path between two nodes.

The disadvantage of this approach is a very expensive path discovery. As the whole network is flooded with a request, this results in poor energy efficiency. Even though other approaches like routing via trees also rely on path discovery, the search in those networks is more directed and does not flood the entire network.

3.2.5. Routing based on routing tables

Approaches based on *routing tables* first execute a learning or a bootstrapping phase during which every node learns routes to nodes that are of interest. For every node, two other pieces of information are usually stored: $\langle distance, next\_hop \rangle$. As a distance, usually number of hops is used, but any other additive metric could be used. Once the bootstrapping phase is complete, each node can independently forward the packet using the locally stored routing table.

Routing algorithms like RPL [23] use a subset of nodes to store the routing table and only for nodes that are in the sub-tree rooted in a given node. Other approaches like *Energy Aware Routing* (EAR) [71] use a routing table to store several alternative routes to the base-station so a node can choose alternative routes in order to better distribute network load. Kolcun et al. proposed Dragon [72], where every node in the network stores a path to every other node in the network. The platform also includes algorithms for quick update of the routing table in the case of a node failure while keeping the network overhead low.

3.3. Dissemination protocols

The purpose of *dissemination* protocols is to deliver information to either all or a subset of the nodes in the network. This type of communication is often referred to as a *one-to-many*

http://dx.doi.org/10.5772/65986

communication pattern. Often the initiator of the communication is the base-station when it wants to, for example, update the reporting interval. Because most of the collection protocols do not support one-to-one communication, dissemination protocols are also often used if a node wants to deliver a message to one particular node only, even at the expense of flooding the whole network.

Two of the basic techniques used are flooding and gossiping [73]. While in the case of flooding each node just re-broadcasts every message it receives, in the case of gossiping, a node upon receiving a message randomly chooses a neighbour to which it forwards the message. The disadvantage of flooding is the implosion and duplication of messages, while the disadvantage of gossiping is a possible large delay in propagation of the message.

Heinzelman et al. proposed a family of *Sensor Protocols for Information via Negotiation (SPIN)* [74]. A node running SPIN, upon receiving a new message, broadcasts an *advertising* message containing meta-data of the received message. Nodes which have not previously received the message reply with a *request* message. The node then broadcasts the *data* message to all the nodes that requested the data. SPIN can significantly decrease the network traffic by eliminating redundant data. However, SPIN cannot guarantee delivery of the message to all the intended recipients due to message loss and unreliable communication.

Braginsky and Estrin proposed *Rumor Routing (RR)* [75] where nodes that wish to distribute information about a certain event generate special long-lived packets called *agents* which are then gossiped through the network. Every node which receives an agent packet creates a record in its *event table*. Subsequently, if a node is interested in a given event, the node generates a query and uses the gossip protocol to propagate it through the network. Any node that has a record in its event table can forward the query to the node that holds the information about the event. Rumor Routing can significantly decrease the network traffic in cases where the number of events is small and the number of queries is large.

Both SPIN and RR scored the lowest mark for energy efficiency—one out of four—in the survey on routing protocols [56], due to their low packet delivery rate and not being energy aware when disseminating the data.

Levis et al. introduced probably the most popular dissemination protocol called *Trickle* [76]. It uses a 'polite gossip' policy where a node periodically broadcasts the message it wants to propagate, but stays quiet if it hears a message from a neighbour which is identical to its own. Because the time interval after the same message is re-broadcast increases exponentially, each node hears only a small trickle of packets. The algorithm can achieve global dissemination of the message at a very low maintenance cost. Trickle facilitates the DRIP, DIP and DHV dissemination libraries available in TinyOS and is a key component of the CTP and RPL protocols.

Kolcun et al. introduced the *Static Attribute Propagation* algorithm as a part of their Dragon platform [72]. The algorithm eliminates unnecessary re-broadcasts of the message by overhearing the messages of its neighbours. It relies on the knowledge of a list of common neighbours at every node. A node upon receiving a message sets up a random timer. While waiting, the node overhears all its neighbours broadcasting the message. Upon expiration of the random

Functionalities	Protocols
Collection	CTP [44], RPL [23], BCP [52], LEACH [53], HEED [55], PEGASIS [54], LWB [77]
Dissemination	SPIN [74], RR [75], Static Attribute Propagation [72], RPL [23], LWB [77], Trickle [76]
Peer-to-peer	GEAR [62], GPSR [60], Geometric Ad hoc Routing [61], ORPL [64], Innet [66], AODV-BR [69], Dragon [72], LWB [77], Chaos [78]

Table 2. Summary of networking protocols by functionality.

timeout, the node checks whether the set of neighbours which broadcast the message cover all the nodes neighbours. If so, the message is discarded, and otherwise, the node broadcasts the message. This approach can significantly decrease the network traffic, especially in dense networks where the number of common neighbours is high.

3.3.1. Special case: low-power wireless bus

The low-power wireless bus (LWB) is a special case that considers multiple traffic patterns, including one-to-many, many-to-one and many-to-many traffic by exploiting the aforementioned Glossy mechanism to facilitate efficient and reliable floods [77]. It uses time synchronisation to manage access to the bus, where a global communication schedule is maintained (computed online based on immediate traffic) and flooded periodically to nodes (thus avoiding relative clock drift). The authors demonstrate that the LWB is comparable to or outperforms a number of state-of-the-art stacks with regard to many-to-one (i.e. collection) traffic, adapts well to varying traffic volumes, significantly outperforms contemporary approaches in terms of many-to-many, is robust to inference and intermittent node participation, and supports mobile nodes as source or sink network devices (**Table 2**).

4. Full stack implementations

4.1. IPv6 over LR-WPAN

With arrival of IPv6, researchers set about implementing it for sensor networks. This was met with several challenges. Because the main usage domain of IPv6 is Ethernet, to cope with increased Internet traffic, the maximum transmission unit (MTU) was increased from 576 to 1280 bytes, when compared to IPv4. IPv6 addresses are 128-bit long, and the standard IPv6 header size is 40 bytes. This is in strict contrast with the IEEE 802.15.4 standard whose throughput is limited to 250 kbps and the length of the frame to 127 bytes. The standard supports two addresses: short 16-bit and EUI-64 extended addresses. With link headers included, the effective size of the payload could be as small as 81 bytes, which make IPv6 headers seem too large.

In 2007, Mulligan and an Internet Engineering Task Force (IETF) working group published a proposal on how to transfer IPv6 packets in low-rate wireless personal area networks. A new protocol called 6LoWPAN [79] was introduced. The aim of the working group was to define a stateless header compression that would decrease the header size so it can be used with the IEEE 802.15.4 standard. The reduction was achieved by introducing four basic header types:

(*i*) Dispatch Header, (*ii*) Mesh Header, (*iii*) Fragmentation Header and (*iv*) HC1 Header (IPv6 Header Compression Header). 6LoWPAN implements variable-sized headers, where the header size varies from 4 to 13 bytes, depending on what kind of communication is required. The protocol is also prepared to support new types of headers in the future.

When an address needs to be included in the header, 6LoWPAN supports either 16-bit short addresses or full 64-bit addresses. These addresses are then translated to full 128-bit IPv6 addresses by a border router. The border router is a router that enables communication between a WSN and the Internet.

If a node needs to send a packet whose size is larger than the size of the payload of 802.15.4 frame (107 bytes), 6LoWPAN defines a fragmentation header which allows the node to split the original datagram into several packets. The header includes the size of the original datagram as well as the ordering number. Fragmentation is also necessary as the specification of IPv6 requires support of a minimum MTU of 1280 bytes.

6LoWPAN supports two types of routing: (*i*) mesh-under and (*ii*) route-over. In the mesh-under approach, routing is done by the link layer (layer two) using IEEE 802.15.4 frame or 6LoWPAN header. To send a packet to a destination node, EUI 64-bit or 16-bit short addresses are used to forward the packet to the next-hop neighbour, preferably closer to the destination. To complete a single IP hop multiple link layer hops may be required. If an IP packet is fragmented into several fragments, these fragments may travel over different paths. The packet is reassembled at the final destination only. The advantage of mesh-under is that the forwarding nodes do not need to reassemble the whole packet to make the routing decision which lowers the memory requirements. Additionally, because the packets may travel via different paths, mesh-under can increase throughput and lower the congestion of the network. On the other hand, if the destination node is missing, a fragment of the IP packet, the whole packet (i.e. all fragments) needs to be resent.

In the route-over approach, the routing decision is done on the network layer (layer 3) and each node acts as an IP router. Each link/hop is considered to be an IP hop too. If a packet is fragmented, then all fragments are first reassembled on the next hop neighbour and the packet is passed to the network layer. The network layer decides whether the packet should be processed on the node or forwarded to a neighbour. To make this decision, the node has to either store the routing table which maps the destination address to the next-hop address or the packet itself has to contain this information. In route-over approach, each node must have enough memory to reconstruct the packet, and all fragments are routed via one path only. On the other hand, if a fragment is lost during the transmission, the whole IP packet must be resent over one link layer hop only.

Recalling **Figure 1**, a full stack implementation requires some higher and lower layer primitives. At the application layer, the IETF has worked to standardise the so-called Constrained Application Protocol (CoAP), which is essentially a RESTful (Representational State Transfer) protocol that uses a small subset of HTTP commands, and more recently TSCH at the link layer, detailed earlier. To-date, there are no comprehensive evaluations of the energy performance of the entire stack. However, there are several which evaluate snapshots of the stack, or subsets thereof (e.g. by layer), such as in the case of CoAP in [80], where the authors show that CoAP is efficient when implemented over RPL, 6LoWPAN and ContikiMAC (and reiterate that the key efficiencies are to be gained at the link layer), and 6TiSCH [81], where a realistic energy model presented for TSCH demonstrated that under certain conditions, sub-1% duty cycles are demonstrable for real and simulated networks under reasonable traffic loads.

4.2. Composable stacks

Many of the protocols described thus far have open source, modular implementations available in the libraries of the various operating systems. Therefore, they can easily be composed to suit an intended application scenario. We know that performance and appropriate selection depend on application level requirements and statistical properties of the network traffic generated by that application. Therefore, there are very few studies that explore in-depth full-stack implementations on per-application bases. However, there are a number well-documented implementations that comparatively evaluate performance such as in [77], where the authors compare the performance of LWB against Dozer (a highly efficient TDMA-based data collection protocol) [82], CTP+A-MAC and CTP+LPL, under a variety of conditions. CTP+A-MAC, LWB and more recently Chaos [78] are representative state-of-the-art stacks from the research community that rival the standards-based stacks which tend to adopt something very similar to the 6TiSCH approach (**Figure 1**), for example. While many of the protocols described so far have been implemented in the longer-standing operating systems developed in the research community, a number of more recent such operating systems have emerged, such as OpenWSN—https://openwsn.atlassian.net/wiki/, and RIOT—https://www.riot-os.org/#features, which provide implementations of the standardised stack for a variety of recent hardware development platforms.

4.3. Energy analysis

One of the most comprehensive evaluations of a relatively complete stack is presented in [44], where CTP is run on a number of heterogeneous test-beds, over a number of link layers, for a variety of inter-packet intervals (typically determined by the application scenario) and at a variety of radio frequencies on various channels. While the evaluation shows that the combination of beaconing and data path validation used in its design is robust over a variety of physical and link layers, the performance characteristics still do not quite meet those needed for ultra-long-lived, large (i.e. extremely dense) and highly reliable applications. The authors also leave open the question of whether these methods are suitable for distance vector algorithms synonymous with *ad hoc* networking.

Generally speaking, it is extremely difficult to validate or comparatively evaluate the energy performance of a protocol stack relative to another. The use of simulators and non-standard test facilities (e.g. community-known test-beds set-up with arbitrary configurations) contribute to this problem. This could be mitigated against by having a set of standard simulation and test-bed configurations against which to benchmark protocols at each layer, and in combination. This has been recently alluded to in the literature in [83], wherein the authors conclude that there is insufficient knowledge available for a majority of the community when

http://dx.doi.org/10.5772/65986

it comes to trialling experiments on real-world facilities such that they can be trustworthy, reproducible and thus independently verifiable.

5. Conclusion and future directions

Practical, energy-efficient wireless communications protocols have been comprehensively studied and documented in the literature in the preceding decades. We have presented a comprehensive summary review of the state-of-the-art concerning link and routing layer technologies developed during this period suitable for constrained wireless sensor network, IoT and CPS applications.

A great deal is known about their limitations and the trade-offs inherent in their selection and implementation. It is arguable that fundamental performance limits have been reached in the design of link layer technologies for contemporary radio transceivers. For this reason, and understanding the time-varying nature of the wireless medium, routing protocols are being developed by the standards bodies that take into account lower layer parameters in their design (e.g. IEEE 802.15.5). These may include link quality, residual energy and dynamic energy availability in the case of energy harvesting devices.

Devices are incrementally more efficient, and their inter-networking based on link and routing layer technologies is maturing to the point where protocols can confidently be selected where certain performance requirements must be satisfied. We tend to readily trade energy efficiency against reliability and determinism for industrial and high-criticality applications. This is problematic, however, because many potential application scenarios are dismissed as economically infeasible due to high network maintenance costs. Simultaneously, as devices and protocols maximise efficiency, the gap with feasible energy harvesting from devices' ambient environments is reducing. Coupled with other techniques and technologies, like compressive and predictive sensing, ultra-low power wake-up radio circuits and so on, there is an emergent design space—where applications can be holistically co-designed with regard to energy. It is almost certain that such approaches will be investigated in the short to medium term, which will result in the economic feasibility of a range of new connected monitoring and control applications.

Author details

David Boyle*, Roman Kolcun and Eric Yeatman

*Address all correspondence to: david.boyle@imperial.ac.uk

Department of Electrical and Electronic Engineering, Imperial College London, England

References

[1] L. Gu and J.A. Stankovic. Radio-triggered wake-up for wireless sensor networks. *Real-Time Systems*, 29(2–3):157–182, 2005.

[2] S.J. Marinkovic and E.M. Popovici. Nano-power wireless wake-up receiver with serial peripheral interface. *IEEE Journal on Selected Areas in Communications*, 29(8):1641–1647, 2011.

[3] M. Magno, D. Boyle, D. Brunelli, B. O'Flynn, E. Popovici, and L. Benini. Extended wireless monitoring through intelligent hybrid energy supply. *IEEE Transactions on Industrial Electronics*, 61(4):1871–1881, April 2014.

[4] R.G. Baraniuk. Compressive sensing. *IEEE Signal Processing Magazine*, 24(4), 118–121 2007.

[5] R.G. Baraniuk, V. Cevher, M.F. Duarte, and C. Hegde. Model-based compressive sensing. *IEEE Transactions on Information Theory*, 56(4):1982–2001, 2010.

[6] U. Raza, A. Camerra, A.L. Murphy, T. Palpanas, and G.P. Picco. What does model-driven data acquisition really achieve in wireless sensor networks? In *2012 IEEE International Conference on Pervasive Computing and Communications (PerCom)*, pp. 85–94. IEEE, 2012.

[7] W.R. Heinzelman, A. Chandrakasan, and H. Balakrishnan. Energy-efficient communication protocol for wireless microsensor networks. In *2000 Proceedings of the 33rd annual Hawaii International Conference on System Sciences*, 10 pp. IEEE, 2000.

[8] L. Mottola, G.P. Picco, M. Ceriotti, S. Gunçž, and A.L. Murphy. Not all wireless sensor networks are created equal: A comparative study on tunnels. *ACM Transactions on Sensor Networks (TOSN)*, 7(2):15, 2010.

[9] D. Boyle, B. Srbinovski, E. Popovici, and B. O'Flynn. Energy analysis of industrial sensors in novel wireless shm systems. In *Sensors, 2012 IEEE*, pp. 1–4, Oct 2012.

[10] W. Ye, J. Heidemann, and D. Estrin. An energy-efficient mac protocol for wireless sensor networks. In *INFOCOM 2002. Twenty-First Annual Joint Conference of the IEEE Computer and Communications Societies. Proceedings. IEEE*, vol. 3, pp. 1567–1576. IEEE, 2002.

[11] A. Moschitta and I. Neri. Power consumption assessment in wireless sensor networks. *ICT-Energy-Concepts Towards Zero-Power Information and Communication Technology*, 2014.

[12] J. Polastre, J. Hill, and D. Culler. Versatile low power media access for wireless sensor networks. In *Proceedings of the 2nd International Conference on Embedded Networked Sensor Systems*, pp. 95–107. ACM, 2004.

[13] T.S. Rappaport, S. Sun, R. Mayzus, H. Zhao, Y. Azar, K. Wang, G.N. Wong, J.K. Schulz, M. Samimi, and F. Gutierrez. Millimeter wave mobile communications for 5g cellular: It will work! *IEEE Access*, 1:335–349, 2013.

[14] SIG Bluetooth. Bluetooth core specification version 4.0. *Specification of the Bluetooth System*, 2010.

[15] T.S Rappaport et al. *Wireless communications: principles and practice*, vol. 2. Prentice Hall PTR ,New Jersey, 1996.

[16] Texas Instruments. Cc2420: 2.4 ghzieee 802.15. 4/zigbee-ready rf transceiver. *Available at* http://www.ti.com/lit/gpn/cc2420, 53, 2006.

[17] Texas Instruments. Cc2538 powerful wireless microcontroller system-on-chip for 2.4-ghz ieee 802.15. 4, 6lowpan, and zigbee applications. *CC2538 datasheet (April 2015)*, 2015.

[18] IEEE standard for information technology—local and metropolitan area networks—specific requirements—part 15.4: Wireless medium access control (mac) and physical layer (phy) specifications for low rate wireless personal area networks (wpans). *IEEE Std 802.15.4-2006 (Revision of IEEE Std 802.15.4-2003)*, pp. 1–320, Sept 2006.

[19] ZigBee Alliance. Zigbee specification, 2006.

[20] J. Song, S. Han, A.K. Mok, D. Chen, M. Lucas, and M. Nixon. Wirelesshart: Applying wireless technology in real-time industrial process control. In *Real-Time and Embedded Technology and Applications Symposium, 2008. RTAS'08. IEEE*, pp. 377–386. IEEE, 2008.

[21] S. Petersen and S. Carlsen. Wirelesshart versus isa100. 11a: the format war hits the factory floor. *Industrial Electronics Magazine, IEEE*, 5(4):23–34, 2011.

[22] R. Madan, S. Cui, S. Lall, and N. A. Goldsmith. Cross-layer design for lifetime maximization in interference-limited wireless sensor networks. *IEEE Transactions on Wireless Communications*, 5(11):3142–3152, Nov. 2006.

[23] T. Winter, P. Thubert, T. Clausen, J. Hui, R. Kelsey, P. Levis, K. Pister, R. Struik, and J. Vasseur. Rpl: Ipv6 routing protocol for low power and lossy networks, rfc 6550. *IETF ROLL WG, Tech. Rep*, 2012.

[24] C. Bormann, K. Hartke, and Z. Shelby. The Constrained Application Protocol (CoAP). RFC 7252, October 2015.

[25] T. Schmid, R. Shea, Z. Charbiwala, J. Friedman, M.B. Srivastava, and Y.H. Cho. On the interaction of clocks, power, and synchronization in duty-cycled embedded sensor nodes. *ACM Trans. Sen. Netw.*, 7(3):24:1–24:19, October 2010.

[26] E. O'Connell, B. O'Flynn, and D. Boyle. Clocks, latency and energy efficiency in duty cycled, multi-hop wireless sensor networks. In *Advances in Sensors and Interfaces (IWASI), 2013 5th IEEE International Workshop on*, pp. 199–204, June 2013.

[27] P. Dutta, S. Dawson-Haggerty, Y. Chen, C-J. Mike Liang, and A. Terzis. Design and evaluation of a versatile and efficient receiver-initiated link layer for low-power wireless. In *Proceedings of the 8th ACM Conference on Embedded Networked Sensor Systems*, pp. 1–14. ACM, 2010.

[28] D. Moss and P. Levis. Box-macs: Exploiting physical and link layer boundaries in low-power networking. *Computer Systems Laboratory Stanford University*, pp. 116–119, 2008.

[29] J.W. Hui and D.E. Culler. Ip is dead, long live ip for wireless sensor networks. In *Proceedings of the 6th ACM Conference on Embedded Network Sensor Systems*, pp. 15–28. ACM, 2008.

[30] A. Dunkels. The ContikiMAC radio duty cycling protocol. *Technical Report T2011:13*, Swedish Institute of Computer Science. December 2011.

[31] A. El-Hoiydi and J-D. Decotignie. Wisemac: An ultra low power mac protocol for multi-hop wireless sensor networks. In *Algorithmic Aspects of Wireless Sensor Networks*, pp. 18–31. Springer, 2004.

[32] K. Langendoen and A. Meier. Analyzing mac protocols for low data-rate applications. *ACM Transactions on Sensor Networks (TOSN)*, 7(2):19, 2010.

[33] E. O'Connell, B. O'Flynn, and D. Boyle. Energy & reliability optimal mac for wsns. In *2014 10th International Conference on the Design of Reliable Communication Networks (DRCN)*, pp. 1–8, April 2014.

[34] A. Bachir, M. Dohler, T. Watteyne, and K.K. Leung. Mac essentials for wireless sensor networks. *Communications Surveys & Tutorials, IEEE*, 12(2):222–248, 2010.

[35] IEEE standard for local and metropolitan area networks–part 15.4: Low-rate wireless personal area networks (lr-wpans) amendment 1: Mac sublayer. *IEEE Std 802.15.4e-2012 (Amendment to IEEE Std 802.15.4-2011)*, pp. 1–225, April 2012.

[36] M. Ceriotti and A.L. Murphy. A mac contest between lpl (the champion) and reins-mac (the challenger, an anarchic tdma scheduler providing qos). In *Proceedings of the 8th ACM Conference on Embedded Networked Sensor Systems*, pp. 371–372. ACM, 2010.

[37] R. Musaloiu-E., C.J.M. Liang, and A. Terzis. Koala: Ultra-low power data retrieval in wireless sensor networks. In *2008 IPSN '08. International Conference on Information Processing in Sensor Networks*, pp. 421–432, April 2008.

[38] Y. Sun, O. Gurewitz, and D.B. Johnson. Ri-mac: A receiver-initiated asynchronous duty cycle mac protocol for dynamic traffic loads in wireless sensor networks. In *Proceedings of the 6th ACM Conference on Embedded Network Sensor Systems*, SenSys '08, pp. 1–14, New York, NY, USA, 2008. ACM.

[39] D. Carlson and A. Terzis. Flip-mac: A density-adaptive contention-reduction protocol for efficient any-to-one communication. In *2011 International Conference on Distributed Computing in Sensor Systems and Workshops* (*DCOSS*), pp. 1–8. IEEE, 2011.

[40] F. Ferrari, M. Zimmerling, L. Thiele, and O. Saukh. Efficient network flooding and time synchronization with glossy. In *2011 10th International Conference on Information Processing in Sensor Networks (IPSN)*, pp. 73–84. IEEE, 2011.

[41] G. Lu, B. Krishnamachari, and C.S. Raghavendra. Performance evaluation of the IEEE 802.15.4 mac for low-rate low-power wireless networks. In *2004 IEEE International Conference on Performance, Computing, and Communications*, pp. 701–706, 2004.

[42] P. Levis, S. Madden, J. Polastre, R. Szewczyk, K. Whitehouse, A. Woo, D. Gay, J. Hill, M. Welsh, E. Brewer, and D. Culler. Tinyos: An operating system for sensor networks. In W. Weber, J.M. Rabaey, and E. Aarts, editors, *Ambient Intelligence*, pp. 115–148. Springer, Berlin Heidelberg, 2005.

http://dx.doi.org/10.5772/65986

[43] K. Pister and L. Doherty. Tsmp: Time synchronized mesh protocol. *IASTED Distributed Sensor Networks*, pp. 391–398, 2008.

[44] O. Gnawali, R. Fonseca, K. Jamieson, D. Moss, and P. Levis. Collection tree protocol. In *Proceedings of the 7th ACM Conference on Embedded Networked Sensor Systems*, SenSys '09, pp. 1–14, New York, NY, USA, 2009. ACM.

[45] T. Istomin, C. Kiraly, and G.P. Picco. Is rpl ready for actuation? A comparative evaluation in a smart city scenario. In *Wireless Sensor Networks*, pp. 291–299. Springer, 2015.

[46] A. Dunkels, B. Gronvall, and T. Voigt. Contiki—a lightweight and flexible operating system for tiny networked sensors. In *2004 29th Annual IEEE International Conference on Local Computer Networks*, pp. 455–462, Nov 2004.

[47] A. Dunkels, F. Österlind, and Z. He. An adaptive communication architecture for wireless sensor networks. In *SenSys '07*, pp. 335–349, 2007.

[48] S. Madden, M.J. Franklin, J.M. Hellerstein, and W. Hong. The design of an acquisitional query processor for sensor networks. In *Proceedings of the 2003 ACM SIGMOD International Conference on Management of Data*, SIGMOD '03, pp. 491–502, New York, NY, USA, 2003. ACM.

[49] J.G. Ko, S. Dawson-Haggerty, O. Gnawali, D. Culler, and A. Terzis. Evaluating the performance of rpl and 6lowpan in tinyos. In *Workshop on Extending the Internet to Low Power and Lossy Networks (IPSN)*, vol. 80, pp. 85–90, 2011.

[50] J.G. Ko, J. Eriksson, N. Tsiftes, S. Dawson-Haggerty, J-P. Vasseur, M. Durvy, A. Terzis, A. Dunkels, and D. Culler. Industry: Beyond interoperability: Pushing the performance of sensor network ip stacks. In *Proceedings of the 9th ACM Conference on Embedded Networked Sensor Systems*, SenSys '11, pp. 1–11, New York, NY, USA, 2011. ACM.

[51] S. Lindsey, C.S. Raghavendra, and K.M. Sivalingam. Data gathering in sensor networks using the energy delay metric. In *Proceedings of the 15th International Parallel & Distributed Processing Symposium*, IPDPS '01, pp. 188, Washington, DC, USA, 2001. IEEE Computer Society.

[52] S. Moeller, A. Sridharan, B. Krishnamachari, and O. Gnawali. Routing without routes: The backpressure collection protocol. In *Proceedings of the 9th ACM/IEEE International Conference on Information Processing in Sensor Networks*, IPSN '10, pp. 279–290, New York, NY, USA, 2010. ACM.

[53] W.R. Heinzelman, A. Chandrakasan, and H. Balakrishnan. Energy-efficient communication protocol for wireless microsensor networks. In *2000 Proceedings of the 33rd Annual Hawaii International Conference on System Sciences*, 10 pp. vol. 2, Jan 2000.

[54] S. Lindsey and C.S. Raghavendra. Pegasis: Power-efficient gathering in sensor information systems. In *Aerospace Conference Proceedings, 2002. IEEE*, vol. 3, pp. 3–1125–3–1130, 2002.

[55] O. Younis and S. Fahmy. Distributed clustering in ad-hoc sensor networks: a hybrid, energy-efficient approach. In *INFOCOM 2004. Twenty-third Annual Joint Conference of the IEEE Computer and Communications Societies*, vol. 1, pp. 640, March 2004.

[56] A. Murtala Zungeru, L.-M. Ang, and K.P. Seng. Classical and swarm intelligence based routing protocols for wireless sensor networks: A survey and comparison. *Journal of Network and Computer Applications*, 35(5):1508–1536, 2012. Service Delivery Management in Broadband Networks.

[57] P. Bahl and V.N. Padmanabhan. Radar: an in-building rf-based user location and tracking system. In *INFOCOM 2000. Nineteenth Annual Joint Conference of the IEEE Computer and Communications Societies. Proceedings. IEEE*, pp. 775–784 vol. 2, 2000.

[58] N. Bulusu, J. Heidemann, and D. Estrin. Gps-less low-cost outdoor localization for very small devices. *Personal Communications, IEEE*, 7(5):28–34, Oct 2000.

[59] L. Doherty, K.S.J. Pister, and L. El Ghaoui. Convex position estimation in wireless sensor networks. In *INFOCOM 2001. Twentieth Annual Joint Conference of the IEEE Computer and Communications Societies. Proceedings. IEEE*, pp. 1655–1663, vol. 3, 2001.

[60] B. Karp and H.T. Kung. GPSR: greedy perimeter stateless routing for wireless networks. In *Proceedings of the 6th Annual International Conference on Mobile Computing and Networking*, MobiCom'00, pp. 243–254, New York, NY, USA, 2000. ACM.

[61] F. Kuhn, R. Wattenhofer, Y. Zhang, and A. Zollinger. Geometric ad-hoc routing: Of theory and practice. In *Proceedings of the Twenty-second Annual Symposium on Principles of Distributed Computing*, PODC '03, pp. 63–72, New York, NY, USA, 2003. ACM.

[62] Y. Yu, R. Govindan, and D. Estrin. Geographical and energy aware routing: A recursive data dissemination protocol for wireless sensor networks. Technical report, Technical report ucla/csd-tr-01-0023, UCLA Computer Science Department, 2001.

[63] K. Iwanicki and M. van Steen. On hierarchical routing in wireless sensor networks. In *Proceedings of the 2009 International Conference on Information Processing in Sensor Networks*, IPSN '09, pp. 133–144, Washington, DC, USA, 2009. IEEE Computer Society.

[64] S. Duquennoy, O. Landsiedel, and T. Voigt. Let the tree bloom: Scalable opportunistic routing with orpl. In *Proceedings of the 11th ACM Conference on Embedded Networked Sensor Systems*, SenSys '13, pp. 2:1–2:14, New York, NY, USA, 2013. ACM.

[65] Burton H. Bloom. Space/time trade-offs in hash coding with allowable errors. *Communications of the ACM*, 13(7):422–426, July 1970.

[66] S.R. Mihaylov, M. Jacob, Z.G. Ives, and S. Guha. A substrate for in-network sensor data integration. In *Proceedings of the 5th Workshop on Data Management for Sensor Networks*, DMSN '08, pp. 35–41, New York, NY, USA, 2008. ACM.

[67] E. Cañete, M. Daz, L. Llopis, and B. Rubio. Hero: A hierarchical, efficient and reliable routing protocol for wireless sensor and actor networks. *Computer Communications*, 35(11):1392–1409, June 2012.

[68] D.B. Johnson and D.A. Maltz. Dynamic source routing in ad hoc wireless networks. In T. Imielinski and H.F. Korth, editors, *Mobile Computing, The Kluwer International Series in Engineering and Computer Science*, vol. 353, pp. 153–181. Springer, USA, 1996.

[69] S-J. Lee and M. Gerla. Aodv-br: backup routing in ad hoc networks. In *Wireless Communications and Networking Conference, 2000. WCNC. 2000 IEEE*, vol. 3, pp. 1311–1316, 2000.

[70] C.E. Perkins and E.M. Royer. Ad-hoc on-demand distance vector routing. In *WMCSA '99. Second IEEE Workshop on Mobile Computing Systems and Applications, 1999. Proceedings*, WMCSA '99, pp. 90–100, Feb 1999.

[71] R.C. Shah and J.M. Rabaey. Energy aware routing for low energy ad hoc sensor networks. In *Wireless Communications and Networking Conference, 2002. WCNC2002. 2002 IEEE*, vol. 1, pp. 350–355, Mar 2002.

[72] R. Kolcun, D.E. Boyle, and J.A. McCann. Efficient distributed query processing. *IEEE Transactions on Automation Science and Engineering*, pp. 1–17, July 2016.

[73] S.M. Hedetniemi, S.T. Hedetniemi, and A.L. Liestman. A survey of gossiping and broadcasting in communication networks. *Networks*, 18(4):319–349, 1988.

[74] W.R. Heinzelman, J. Kulik, and H. Balakrishnan. Adaptive protocols for information dissemination in wireless sensor networks. In *Proceedings of the 5th Annual ACM/IEEE International Conference on Mobile Computing and Networking*, MobiCom '99, pp. 174–185, New York, NY, USA, 1999. ACM.

[75] D. Braginsky and D. Estrin. Rumor routing algorithm for sensor networks. In *Proceedings of the 1st ACM International Workshop on Wireless Sensor Networks and Applications*, WSNA '02, pp. 22–31, New York, NY, USA, 2002. ACM.

[76] P. Levis, N. Patel, D. Culler, and S. Shenker. Trickle: A self-regulating algorithm for code propagation and maintenance in wireless sensor networks. In *Proceedings of the 1st Conference on Symposium on Networked Systems Design and Implementation - Volume 1*, NSDI'04, pp. 2–2, Berkeley, CA, USA, 2004. USENIX Association.

[77] F. Ferrari, M. Zimmerling, L. Mottola, and L. Thiele. Low-power wireless bus. In *Proceedings of the 10th ACM Conference on Embedded Network Sensor Systems*, SenSys '12, pp. 1–14, New York, NY, USA, 2012. ACM.

[78] O. Landsiedel, F. Ferrari, and M. Zimmerling. Chaos: Versatile and efficient all-to-all data sharing and in-network processing at scale. In *Proceedings of the 11th ACM Conference on Embedded Networked Sensor Systems*, SenSys '13, pp. 1:1–1:14, New York, NY, USA, 2013. ACM.

[79] G. Mulligan. The 6lowpan architecture. In *Proceedings of the 4th Workshop on Embedded Networked Sensors*, EmNets '07, pp. 78–82, New York, NY, USA, 2007. ACM.

[80] M. Kovatsch, S. Duquennoy, and A. Dunkels. A low-power coap for contiki. In *2011 IEEE Eighth International Conference on Mobile Ad-Hoc and Sensor Systems*, pp. 855–860, Oct 2011.

[81] X. Vilajosana, Q. Wang, F. Chraim, T. Watteyne, T. Chang, and K.S.J. Pister. A realistic energy consumption model for TSCH networks. *IEEE Sensors Journal*, 14(2):482–489, Feb 2014.

[82] N. Burri, P. von Rickenbach, and R. Wattenhofer. Dozer: Ultra-low power data gathering in sensor networks. In *Proceedings of the 6th International Conference on Information Processing in Sensor Networks*, IPSN '07, pp. 450–459, New York, NY, USA, 2007. ACM.

[83] G. Z. Papadopoulos, K. Kritsis, A. Gallais, P. Chatzimisios, and T. Noel. Performance evaluation methods in ad hoc and wireless sensor networks: a literature study. *IEEE Communications Magazine*, 54(1):122–128, January 2016.

[84] P. Thubert. An architecture for IPv6 over the TSCH mode of IEEE 802.15.4. Internet-Draft draft-ietf-6tisch-architecture-09, Internet Engineering Task Force, November 2015. Work in Progress.

Chapter 8

Globally Optimised Energy-Efficient Data Centres

Dirk Pesch, Susan Rea, J. Ignacio Torrens,
Vojtech Zavrel, J.L.M. Hensen, Diarmuid Grimes,
Barry O'Sullivan, Thomas Scherer, Robert Birke,
Lydia Chen, Ton Engbersen, Lara Lopez, Enric Pages,
Deepak Mehta, Jacinta Townley and
Vassilios Tsachouridis

Additional information is available at the end of the chapter

http://dx.doi.org/10.5772/65988

Abstract

Data centres are part of today's critical information and communication infrastructure, and the majority of business transactions as well as much of our digital life now depend on them. At the same time, data centres are large primary energy consumers, with energy consumed by IT and server room air conditioning equipment and also by general building facilities. In many data centres, IT equipment energy and cooling energy requirements are not always coordinated, so energy consumption is not optimised. Most data centres lack an integrated energy management system that jointly optimises and controls all its energy consuming equipments in order to reduce energy consumption and increase the usage of local renewable energy sources. In this chapter, the authors discuss the challenges of coordinated energy management in data centres and present a novel scalable, integrated energy management system architecture for data centre wide optimisation. A prototype of the system has been implemented, including joint workload and thermal management algorithms. The control algorithms are evaluated in an accurate simulation-based model of a real data centre. Results show significant energy savings potential, in some cases up to 40%, by integrating workload and thermal management.

Keywords: energy efficient data centres, workload management, thermal management, integrated data centre energy management platform

1. Introduction

Data centres have become a critical part of modern information technology (IT) infrastructure with software as a service, mobile cloud applications, digital media streaming and the expected growth in the Internet of Everything all relying on data centres. However, data centres are also significant primary energy users and now consume in the order to 3% of worldwide electricity and are responsible for 2% of global greenhouse gas emissions, the same as the airline industry [1]. With the increasing move towards cloud computing and storage as well as everything as a service type computing, data centre energy consumption is currently growing at a compound annual rate of over 10% and expect to reach approximately 8% of global energy consumption by 2020 [2, 3]. While the hyper-scale data centres of large cloud service providers are consuming in the 10 s of megawatts of power with corresponding annual electricity bills in the order of tens of millions of dollars, for example, Google with over 260 MW and $67 M and Microsoft with over 150 MW and $36 M in 2010 [4], those large cloud service providers are also investing heavily in energy efficiency and green data centres, for example, Google and Microsoft have invested over $900 M in energy reduction measures since 2010. However, smaller operators, independent and co-location/multi-tenant data centres have not yet been able to deploy many of the energy efficiency technologies that are available. This is due to lack of integrated technology solutions and uncertainty about costs and the use of renewable energy solutions. In particular, the many server rooms and small data centres run by commercial businesses and universities are the dominant electricity users as shown in **Figure 1** [5].

On average, computing consumes 60% of total energy in data centres while cooling consumes 35% [6]. New server and cooling technologies have the potential to lead to a 40% reduction of energy consumption, but computation and cooling typically operate without joint coordination or optimisation. While server energy management can reduce energy use at CPU, rack and overall data centre level, dynamic computation scheduling is often neither efficient with many idle servers running rather than being shutdown [5] nor is it generally integrated with cooling. Data centre cooling typically operates at constant cold air temperature to protect the hottest server racks, while local fans distribute the air across racks. However, these local

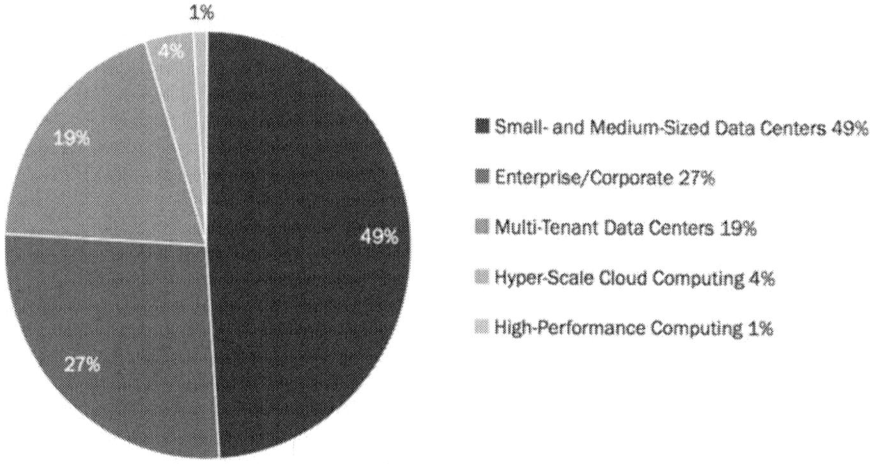

Figure 1. Estimated US data centre electricity consumption by market segment (2011) [5].

http://dx.doi.org/10.5772/65988

server controls are typically not integrated with room cooling systems, which means that it is not possible to optimise chillers, air fans and server fans as a single, whole system.

In order to reduce the CO_2 footprint of data centres, large organisations such as Google and Facebook are investing in renewable energy sources (RES), such as solar photovoltaics (PV) or wind power, often co-located with their hyper-scale data centres [7, 8]. However, for the many smaller data centres and server rooms, the use or integration of renewable energy sources has received limited interest. The reason for this is that these data centres are typically embedded in buildings that also hold other functions, for example, office and meeting spaces, laboratories and lecture rooms in the case of universities. A major issue in this is also the lack of interoperability of generation, storage and heat recovery and current installation and maintenance costs versus payback [9]. By and large, data centre operators, who want to be green and use renewable energy, buy electricity that has been given a green label by their respective supplier without often being able to fully verify this. The intermittency of renewable energy generation is also a critical factor in an environment with very strict service level agreements and essentially 100% uptime requirements. The adoption of new technologies related to computing, cooling, generation, energy storage and waste heat recovery individually requires sophisticated controls, but no single manufacturer provides a complete system, so integration between control systems does not exist.

However, research has been under way in a cluster of projects funded by the European Commission's Framework Programme for Research and Innovation. The cluster includes projects such as DC4Cities, GENiC, CoolEmAll, RenewIT, Eureca, GEYSER, GreenDataNet, Dolfin and All4Green, which are all focused on a range of aspects to increase data centre energy efficiency but also to integrate data centre energy use and recovery into a future smart grid and smart city environment. One of those projects, GENiC (http://www.projectgenic.eu), in particular, aims at developing integrated cooling and computing control strategies in conjunction with innovative power management concepts that incorporate renewable electrical power supply and storage, and waste heat management. The project's aim is to address the issue mentioned above by developing an integrated, flexible, component-based management and control platform for data centre wide optimisation of energy consumption, reduction of carbon emissions and increased local renewable energy supply usage through integrating monitoring and control of computation, data storage, cooling, on-site power generation and waste heat recovery.

A key element in not only achieving a reduction in energy consumption but also a reduction in carbon emissions is energy supply by renewable energy generation and, where possible, energy storage equipment. Such an approach needs to be operated as a complete system to achieve an optimal energy and emissions outcome. This vision of integrated, holistic energy management is centred on the development of a hierarchical control system to operate all of the primary data centre components in an optimal and coordinated manner.

2. Challenges for integrated data centre energy management

While data centres have become a critical IT infrastructure and also a significant consumer of energy and contributor to CO_2 emissions, opportunities exist to enhance the energy and power management of data centres in conjunction with renewable energy generation and integration with their surrounding infrastructure. Work has been done on studying the topic of powering of data centres by renewable energy [10], but this has not been fully integrated into a complete energy management system considering coordinated workload management, cooling, powering and heat recovery management. While much work has focused on integrated energy management for data centres [11, 12], there is still a lack of an overall consideration of energy usage and powering with the recovery of waste heat as part of an overall thermal management approach. In order to bring the elements of workload management, cooling, powering and heat recovery together in such a way that it will be possible to achieve a high level of renewable energy powering of data centres, a comprehensive integrated energy management system is needed. The challenges that such a system needs to address are as follows:

- **Comprehensive, scalable integration of workload management with cooling approaches:** in most data centres, workload is allocated to servers without consideration of the thermal impact that this has on the data centre space. In many cases, idle servers are not even shutdown and continue to consume energy without any productive IT load processing. An integration of IT workload management with cooling through thermally aware workload consolidation is required.

- **Effective power management with a high level of renewable energy supply integration while meeting service level agreements:** in order to facilitate the uptake in renewable energy supply systems, in particular at a local level, intelligent power management approaches are needed to balance the intermittently available renewable energy sources, for example, solar, wind, with grid supplied electricity while managing service level agreements. Power management needs to also take energy price fluctuations and demand response requirements into consideration to maximise the cost-effectiveness of renewable power solutions in order to create incentives for investment in such solutions.

- **Strategies for waste heat recovery in conjunction with the heating needs of surrounding areas:** opportunities exist for small- to medium-sized data centres to reuse the heat generated by IT workload in order to heat adjacent spaces rather than dump the heat into the air through heat exchangers or dry coolers. Heat recovery solutions can heat spaces or water either within the same building or for larger data centres spaces in adjacent buildings or feed heat into local district heating systems. In this way, heat recovery can reduce the energy demands of adjacent facilities and achieve an overall reduction of energy consumption within the area of the data centre.

- **Design and decision support tools assisting data centre operators with data centre energy management:** for many data centres, in particular for small- to medium-sized data centres embedded into larger organisations, the IT manager and the facilities manager are different roles and as such do not have complete understanding of the complete energy management needs and opportunities. As such, suitable tools are required to assist operators with decision-making in terms of what energy management approaches, power solutions or heat recovery techniques might be most suitable for their situation.

- **Effective monitoring and fault management:** maintain service level agreements and uptime is of paramount importance to data centre operators, above and beyond of everything else.

In order to achieve this while making sure energy consumption costs do not exceed certain levels, effective monitoring and fault management tools are important and can assist operators with their work.

3. An architecture for globally optimised energy management in data centre

To address the challenges outlined above, the EC-funded GENiC project has developed a high-level architecture for an integrated design, management and control platform, targeting data centre wide optimisation of energy consumption by encapsulating monitoring and control of IT workload, data centre cooling, local power generation, energy storage and waste heat recovery. The developed management platform includes control and optimisation, decision support, and fault detection functions and defines interfaces and common data formats to enable a component-based design. The GENiC architecture can act as a template for a wide range of implementations of data centre energy management systems suited to a particular data centre configuration. In the following, a functional specification of the GENiC architecture is presented and an overview of the integration framework is provided. The applicability of the proposed functional architecture is illustrated by a number of use cases. More detail can be found in [13].

3.1. Functional architecture

The GENiC architecture integrates workload management, thermal management and power management by using a hierarchical control concept that enables the coordination of the management sub-systems in an optimal manner with respect to the cost of energy consumption, environmental impact and cost policies. **Figure 2** provides an overview of the developed GENiC system architecture, which consists of six functional groups, the GENiC component groups (GCGs):

- The **Workload Management GCG** is responsible for monitoring, analyzing, predicting, allocating and actuating IT workload within the data centre.
- The **Thermal Management GCG** is responsible for monitoring the thermal environment and cooling systems in the data centre, predicting temperature profiles and cooling demand, and optimally coordinating and actuating the cooling systems.
- The **Power & RES Management GCG** is responsible for monitoring and predicting power supply and demand, and for actuating the on-site power supply of the data centre.
- The **Supervision GCG** includes the supervisory intelligence which provides policies to the power, thermal and workload GCGs for supplying electrical power to meet the IT and cooling power demands of a DC based on monitoring data, predicted systems states and actuation feedback.
- The **Support Tools GCG** includes a number of tools that provide decision support for data centre planners, system integrators and data centre operators.

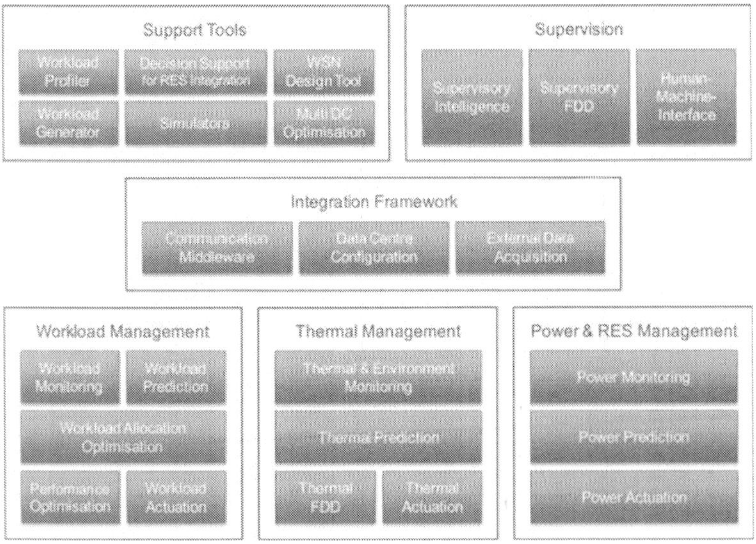

Figure 2. Overview of the GENiC architecture (from [13]).

- The **Integration Framework GCG** provides the communication infrastructure and data formats that are used for interactions between all components of the GENiC system.

Each GCG is composed of a number of functional components, the GENiC components (GCs) (see **Figure 2**). The core function of the GENiC system for continuous data centre energy optimisation can be divided into four basic steps:

1. **Monitoring** components within the management GCGs collect data about IT workload, thermal environment, cooling systems, power demand and on-site power supply.
2. **Prediction** components within the management GCGs update their internal models and estimate future system states based on the collected monitoring data.
3. **Optimisation** components determine optimal policies based on the collected monitoring data and calculated prediction data. These policies are provided to the management GCGs.
4. **Actuation** components within the individual management GCGs implement the policies provided by the optimisation components in the data centre and at the renewable energy sources facilities.

These elements are complemented by components for external data acquisition and fault detection and diagnostics. The basic information flow for coordinating workload, thermal and power management is illustrated in **Figure 3**. In the following, the GENiC component groups are described in more detail.

http://dx.doi.org/10.5772/65988

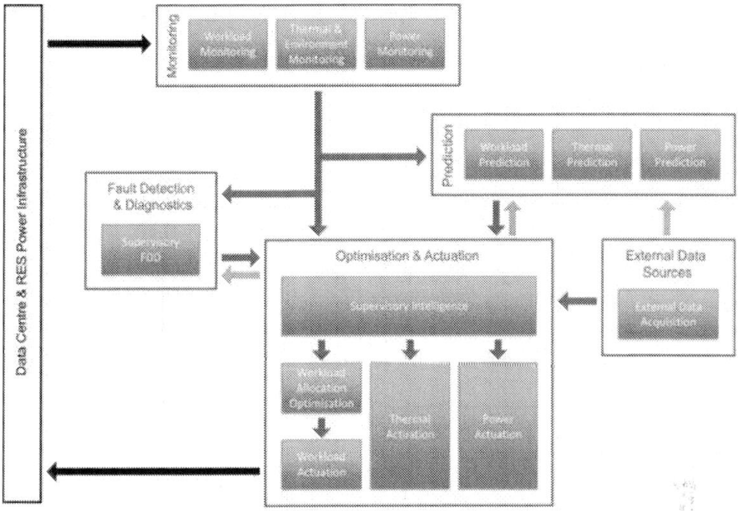

Figure 3. Information flow (simplified) for coordinating workload, thermal and power management [14].

Workload Management GCG: The primary objective of this GCG is to allocate virtual machines (VMs) to physical machines (PMs) such that service level objectives (SLOs) are satisfied with low operational cost. Monitoring data from the IT resources deployed within the data centre are collected by the Workload Monitoring GC. The Workload Prediction GC uses this information to provide short- and long-term predictions on resource utilization. The allocation and migration of VMs to PMs are determined by the Workload Allocation Optimisation GC, which solves a constrained optimisation problem, taking the predicted workload as well as constraints provided by the Supervisory Intelligence GC, Thermal Prediction and Performance Optimisation GC into consideration. The Performance Optimisation GC defines location constraints for individual VMs and modifies the individual VMs' priorities to fulfil application specific SLOs. The VM allocation plan is finally applied by the Workload Actuation GC, which provides an interface to the data centre-specific virtualization platform.

Thermal Management GCG: The Thermal & Environment Monitoring GC integrates monitoring of cooling systems and a sensor network infrastructure for collecting temperature and other environmental data in the data centre space. The collected data are used by the Thermal Prediction GC to provide short-term and long-term predictions to support supervisory control decisions, thermal actuation and workload allocation. Long-term predictions are used for making decisions at the supervisory level. Short-term thermal predictions are required by the Thermal Actuation GC along with real-time sensor measurements to determine optimal set points for the cooling system in order to achieve the targets set by the Supervisory Intelligence GC. These short-term thermal predictions are also necessary input to the Workload Allocation Optimisation GC, as they include temperature models for the thermal contribution of IT server workload to the server inlets and the Supervisory Intelligence GC. Furthermore, short-term predictions, combined with equipment fault information from the Thermal Fault Detection & Diagnostics (FDD) GC, are used for fault detection and diagnostics at the supervisory level.

Power & RES Management GCG: The Power Monitoring GC provides power monitoring information of the DC (power consumed per server, per rack level and total DC power demand), as well as integrates monitoring of the RES infrastructure for local energy generation and storage with data centre power consumption requirements. These data are used by the Power Prediction GC to provide IT Power prediction as well as long-term predictions to support supervisory control decisions and power actuation. The Power Actuation GC determines operation set points for the power systems based on operation policies provided by the Supervisory Intelligence GC and adjusting them depending on measured data and operational conditions.

Supervision GCG: The Supervisory Intelligence GC is responsible for the overall coordination of workload, thermal, power management and heat recovery. It considers power demand and supply, grid energy price, energy storage availability and determines how much power should be supplied from the electricity grid, RES and energy storage to achieve a particular objective on power usage. To this end, it provides policies for the components in the Workload Management, Thermal Management and Power & RES Management GCGs based on information from monitoring and prediction components. The Supervisory Intelligence GC provides these high-level policies for the purpose of guiding the individual management functions towards the Supervisory Intelligence objective strategy that has been chosen as the driver for current data centre operations. Key objective choices might be minimization of financial cost, minimization of carbon emissions or maximization of RES usage. To detect and diagnose system anomalies, the Supervisory FDD GC compares predicted values with measurement data and collects and evaluates fault information. In appropriate situations, the Supervisory FDD GC informs the Supervisory Intelligence GC when a deviation becomes substantial enough to negatively impact system operation so that mitigation action can be taken by the platform until the fault has been corrected. The Human-Machine Interface GC provides a framework for user interfaces that allow data centre operators to monitor and evaluate aggregated data provided by the individual GCs.

Integration Framework GCG: The Communication Middleware GC provides the communication infrastructure used within the GENiC platform. The Data Centre Configuration GC uses a centralized data repository to store all information related to the data centre configuration, including information on data centre layout, cooling equipment, monitoring infrastructure, IT equipment and virtual machines running in the data centre. Finally, the External Data Acquisition GC provides access to data not collected by existing components of the GENiC platform, including weather data, grid energy prices and grid energy CO_2 indicators.

The GENiC platform integrates distributed software components, which are developed and maintained by individual consortium partners. A software component can implement a single

http://dx.doi.org/10.5772/65988

GC, multiple GCs or just part of a GC to provide the required functionality to the platform. For example, a topic-based publish-subscribe messaging architecture is a suitable mechanism to ensure a robust data exchange between individual software components. With this approach, the components do not need to be connected directly to each other, but components can publish messages to a central message broker using pre-defined topics and subscribe to the broker to topics from other components that are of interest to them. The broker forwards all incoming messages to the appropriate subscribers. The GENiC architecture defines a consistent interface specification using a common data format for all GENiC components. All interfaces are defined by hierarchically structured topics. Each of these topics has a defined message payload structure that uses the GENiC common data exchange format which is specified based on JSON [15]. This approach creates a very flexible data centre management platform that can be configured to suit individual, local data centre configurations.

Support Tools GCG: The GENiC platform includes a number of tools to assist data centre planners, system integrators and data centre operators:

- The Workload Profiler GC consists of a set of tools to capture application profiles that can be used by data centre operators to improve application performance.
- The Decision Support for RES Integration GC is a tool for data centre planners to determine the most cost-efficient renewable energy systems to install at a data centre facility.
- The Wireless Sensor Network (WSN) Design Tool GC is a tool to capture system and application level requirements for data centre wireless monitoring infrastructure deployments.
- The Workload Generator GC provides recorded and synthetic VM resource utilization traces for the simulation-based assessment of a GENiC-based system and its implemented algorithms and policies.
- The Simulator GC supports the testing of individual and groups of GCs as well as the (virtual) commissioning of a GENiC platform before its deployment in an actual data centre.
- The Multi Data Centre (DC) Optimisation GC is a tool that exploits the differences in time-zones, energy tariff plans, outside temperatures, performances of geographically distributed data centres to allocate workload amongst them in order to minimise global energy cost and related metrics.

3.2. Energy management use case

The GENiC project's focus to optimally operate data centres with respect to energy is achieved through the integration of workload management, thermal management and power management (including powering through renewable energy sources) via a hierarchical supervisory control concept. Key optimisation criteria in consideration by data centre operators are (i) meeting agreed service level agreements (SLAs), (ii) minimisation of total energy costs, and (iii) with the availability of renewable energy sources also, the maximisation of RES power use and minimisation of carbon emissions. To account for fluctuations in the IT workload demand and the availability of renewable energy supply (which includes local on-site energy production and grid power), the set points of the management sub-systems have to be adapted over time. The Supervisory Intelligence (SI) GC coordinates the individual management sub-systems, including renewable energy supply, by providing optimal policies with respect to the selected optimisation criterion. The use case scenario is illustrated in **Figure 4**. The basic operational flow is as follows [14]:

Step 1—The monitoring GCs, Workload Monitoring, Thermal & Environment Monitoring, and Power Monitoring, collect data from VMs, PMs, air conditioning equipment, sensor networks, power meters and on-site energy supply systems. The relevant information is forwarded to the individual prediction and actuation GCs and SI.

Step 2—Based on recent and historical monitoring data, the prediction GCs, Workload Prediction, Thermal Prediction, and Power Prediction, predict server power demand, thermal profile and cooling demand, RES production capacity and energy demand. The relevant information is forwarded to the individual actuation GCs and SI.

Step 3—Additional data, that is, weather data and grid energy prices, are obtained from external data sources and forwarded to SI by the External Data Acquisition GC.

Step 4—SI provides a set of policies to the actuation GCs, Workload Allocation Optimisation, Thermal Actuation and Power Actuation that are based on inputs from the monitoring and prediction components and further interactions with the Power Prediction GC. These interactions validate the consequences of particular power profiles that SI considers as part of the policy definition. The Workload Allocation Optimisation GC solves a constrained optimisation

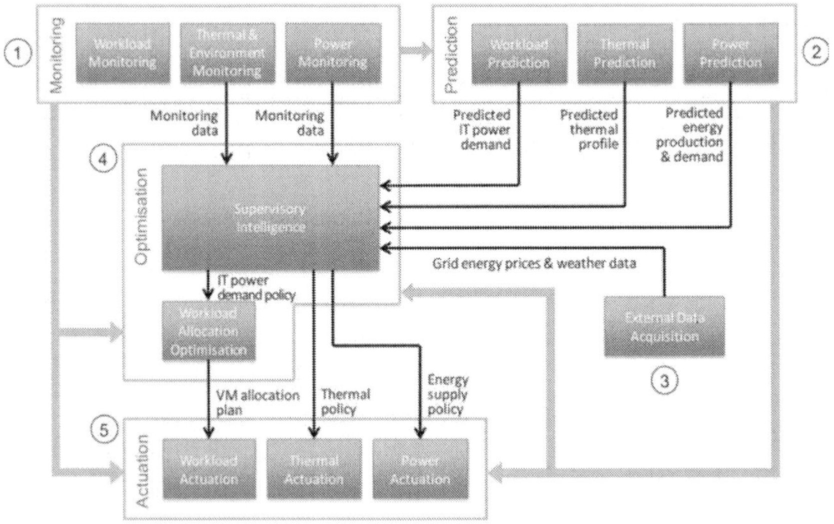

Figure 4. Energy management use case [14].

http://dx.doi.org/10.5772/65988

problem to determine an optimal VM allocation plan minimizing server energy consumption, taking the upper-bound IT power budget recommended by SI and additional inputs from other GCs (thermal and colocation and anti-colocation constraints) into consideration. The Thermal Actuation GC takes the minimum and maximum allowable data centre temperatures determined and then provided to it by SI and optimally calculates cooling equipment set points that ensure the room's thermal profile is properly regulated with minimal cooling equipment electrical power consumption. The Power Actuation GC implements the distribution plan for drawing electricity from grid, controllable and uncontrollable RES, and the schedule for charging and discharging the energy storage device.

Step 5—Based on the inputs from SI and the Workload Allocation Optimisation GC, as well as monitoring and prediction components, the actuation GCs, Workload Actuation, Thermal Actuation, and Power Actuation, decide and apply the actual control actions. For example, the Workload Actuation GC executes the VM allocation plan and switches PMs on/off, based on the actuation requests. Faults are reported back to the optimisation GCs to be considered in the next iteration of the optimisation process.

4. Prototype implementation

Figure 5 illustrates a prototype implementation of the GENiC architecture. The GENiC distributed architecture approach with clearly defined interfaces simplifies integration of a diverse set of software components and allows flexible configuration of the platform. Due to the diverse set of technologies in use in data centres, for example, IT systems, cooling systems, power systems and RES facilities, there is typically no individual manufacturer who supplies all the systems that a data centre requires. Therefore, a data centre management system architecture needs to allow for the integration of individual components supplied by multiple manufacturers and service providers. The architecture detailed in Section 3 is scalable and flexible at the same time and is based on micro-service architecture principles that offer the following benefits:

- **Separation of concerns**—each service implements a single operational functionality. The architecture becomes more flexible and scalable at the same time.
- **Distributed security compliance**—each service can have different security policies, allowing each service provider to maintain local security policies.
- **Freedom of service implementation**—each service provider can choose any development language without compromising the integrity of the overall platform. The only requirement is that the service needs to be able to communicate with the messaging broker.
- **Service scalability**—new instances of services can be spawned when more processing power is required.
- **Simplified API**—all modules use a common API to exchange data and trigger events used by other services.

- **Simplified testing and integration**—testing and integration are easier as testing focuses on black box testing with implementation details hidden behind APIs. Service integration hides APIs and dependencies.

A central element of the implementation of the prototype is the use of the RabbitMQ messaging system [16] for the exchange broker. RabbitMQ provides a range of client implementations in a wide range of programming languages, which allows manufacturers to suit their individual technology set-ups. A Generic Client architecture has been developed to allow each component provider expose their components in a distributed manner in the architecture. The individual GENiC components are implemented as services that communicate via the message broker. The client architecture also offers an easy way to integrate 3rd party (closed source) services with a minimal effort. Each of the components implemented in the GENiC prototype are shown in **Figure 5**, colour coded based on the component group they belong to. Short-term monitored data are stored in a database backend in the GENiC prototype implementation. CouchDB as a NoSQL solution is used, but many other data base solutions are possible depending on the specific needs and data volumes of a particular configuration. Due to the large quantity of stored data, only short-term data are available on the broker.

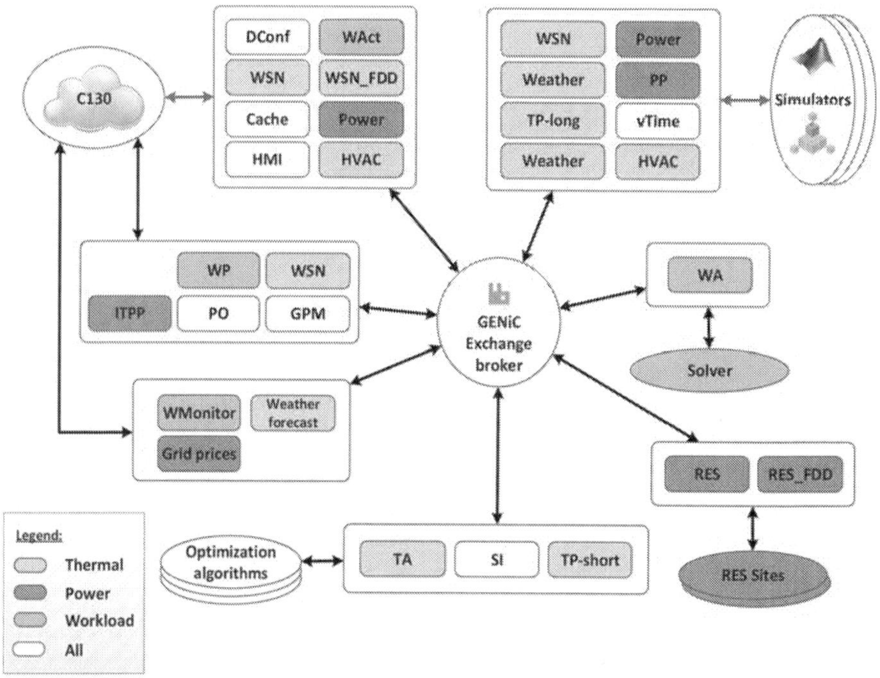

Figure 5. GENiC architecture implementation prototype.

http://dx.doi.org/10.5772/65988

5. Assessment of energy efficiency

In order to assess the effectiveness of data centre management systems in terms of the energy efficiency, power management, managing increased penetration of renewable energy sources, heat reuse and data centre flexibility, the need to select appropriate metrics is of paramount importance. The aforementioned cluster of European research projects on data centre energy efficiency has taken five common data centre metrics and defined 21 new metrics, along with measurement methodologies, to adequately capture the energy efficiency, flexibility and sustainability of modern data centres [17]. This approach supports the development of a common framework for monitoring and assessing the flexibility and sustainability of data centres. The metrics of specific interest for the evaluation of an integrated energy management platform, which integrates thermal and workload management with renewable energy/power supply and heat recovery, are listed in **Table 1**.

The GENiC project considers two types of evaluation: one is based on simulation-based assessment (SBA), which uses the Simulators GENiC component (see **Figure 2**), provided by the tools that have been developed in the project. The Simulators component provides a virtual data centre based on TRNSYS model implementation and simulation and additional interfacing and timing functions [18]. The SBA uses the full energy management platform in the same manner as it is used in a real physical data centre. SBA has the advantage that a specific architecture configuration can be tuned to a particular data centre set-up before deployment in the real environment. This allows for a priori energy efficiency assessment, which not only enables data centre operators to understand what energy savings can be expected from a deployment of an integrated data centre energy and power management platform, but also prepares the platform to run optimally once deployed without affecting the real environment during an in situ tuning process.

Metric	Goal
PUE—Power Usage Effectiveness	Energy/Power Consumption
CER—Cooling Effectiveness Rate	
CUE—Carbon Usage Effectiveness	
Energy Effectiveness of Cooling Mode in a Season	
ERE—Energy Reuse Effectiveness	Energy Recovered/Heat Recovered
APCren—Adaptation of Data Centre to Available Renewable Energy	Data Centre Flexibility—Energy Shifting
DCA—Change in Data Centre Energy Profile from Baseline	
RenPercent—Share of Renewables in Data Centre Electricity Consumption	Renewables Integration
Renewable Energy Factor	
CO_2 Savings Change in Data Centre CO_2 Emissions From Baseline	Primary Energy Savings and CO_2 avoided emissions

Table 1. GENiC evaluation metrics.

The second evaluation is based on the deployment of the prototype in a real data centre. The project chose a small but typical data centre at Cork Institute of Technology. The data centre was adapted to the needs of the project to enable extensive control of the thermal management side, including heat recovery and both virtualisation of the computing infrastructure and normal operation. Experimental renewable energy facilities are linked in a virtual manner to the data centre as the renewable energy micro-grids are located on two premises of project partner Acciona in Spain. The demonstration of use of renewable energy is possible by recording the amount of energy that can be generated by typical micro-grids over time and accounting the amount of electricity flowing into the data centre as either non-renewable or renewable.

5.1. Simulation model—virtual C130 data centre

In order to evaluate the performance of the GENiC platform and to allow pre-deployment assessment and tuning, the project has developed a Simulators GC, which is part of the Support Tools GCG. The simulator component includes energy models that emulate the performance of a data centre and its systems, supporting the development and testing of GENiC components as well as the commissioning of the overall GENiC platform, prior to its physical deployment to the real data centre [19]. The Simulators GC consists of energy models shown in **Figure 6**. These are on the demand side, for example, data centre environment (building energy model and building airflow model), IT devices model, and heating, ventilation and air conditioning (HVAC) systems model, and the supply side, for example, power supply model.

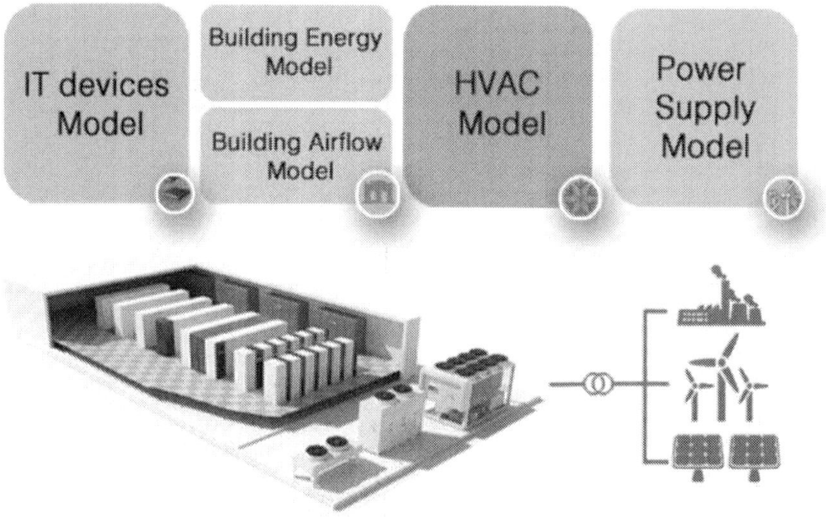

Figure 6. Types of energy models in the Simulator GC.

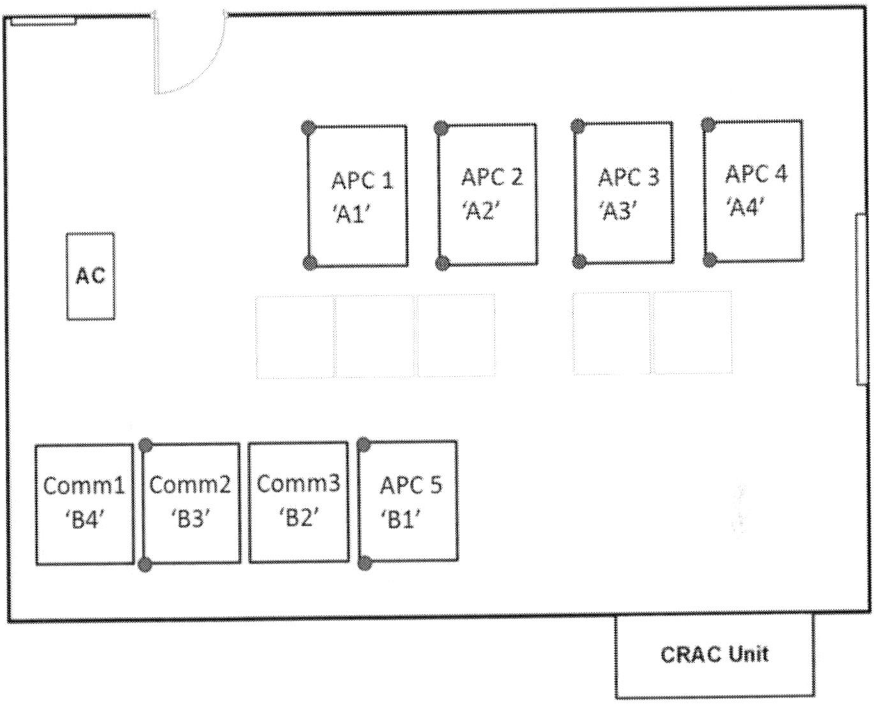

Figure 7. Floor plan of the data centre room used for the simulation-based assessment.

In order to demonstrate the functionality and feasibility of this approach, the Simulator GC implements a virtual data centre model that is based on the actual GENiC demonstration site, the C130 data centre at Cork Institute of Technology. The data centre space is cooled by one main computer room air conditioning unit (CRAC) and one backup air conditioning unit (AC) as illustrated in the floor plan depicted in **Figure 7**.

5.2. IT equipment and DC whitespace characteristics

To emulate the server workload in the data centre, a set of virtual machine (VM) configurations and the VMs' resource utilization traces are required. The traces used for the evaluation example presented here have been collected from a typical corporate data centre production environment and reflect typical enterprise workloads seen in a private cloud environment. The traces comprise resource utilization data for 2400 different VMs hosted on 132 servers. The key parameters of these servers are summarized in **Table 2**. The last column shows the number of servers of each specific type. Each server's dynamic power consumption is modelled as

$$P_{server} = \left(P_{\max} - P_{idle}\right) \times u + P_{idle}$$

where u is the CPU utilization, P_{max} is the server's power consumption at full load (i.e. $u = 1.0$), and P_{idle} is the server's power consumption at idle state (i.e. $u = 0.0$). The total power consumption of the 132 servers is 24.5 kW if all servers operate at full load.

For the simulation-based evaluation example, each server has been mapped to a specific rack space in the simulated data centre. **Table 3** shows this mapping.

Type	CPU size [vcores]	CPU speed [MHz]	Mem. [GB]	Max. power [W]	Idle power [W]	# Servers
S1	8	3200	16	90	30	3
S2	8	3200	32	95	35	8
S3	8	3200	64	105	45	48
S4	12	2000	64	130	70	2
S5	12	2000	128	140	80	12
S6	12	2000	256	160	100	23
S7	24	2700	128	300	140	19
S8	32	2000	128	400	270	14
S9	32	2900	128	460	300	3

Table 2. Server parameters.

Rack	Servers (top to bottom)	ΣP_{max}
B1	2 × S5, 6 × S3, 6 × S6, 6 × S8	4.3 kW
B2	No active equipment; patch panels only	0 kW
B3	10 × S3, 6 × S3, 3 × S6, 4 × S7, 2 × S8	4.2 kW
B4	No active equipment; patch panels only	0 kW
A1	2 × S4, 3 × S1, 8 × S3, 8 × S7, 2 × S5	4.1 kW
A2	4 × S3, 2 × S2, 4 × S5, 5 × S7, 2 × S8, 3 × S3	3.8 kW
A3	4 × S8, 4 × S6, 7 × S3, 4 × S5, 4 × S6	4.2 kW
A4	3 × S9, 6 × S6, 4 × S3, 6 × S2, 2 × S7	3.9 kW

Table 3. Mapping of servers to racks in the virtual data centre.

5.3. Cooling system characteristics

The environment of the data centre is maintained at temperatures between 18 and 27°C with a relative humidity of 30–60% as recommended by ASHRAE [20]. The CRAC unit ensures the required indoor climate. Supply air is distributed through a raised floor and goes to front side of IT devices through perforated tiles. Return air is drawn by the CRAC unit below the ceiling as shown in **Figure 8**.

http://dx.doi.org/10.5772/65988

The conditions of circulating air are controlled in the CRAC unit by a direct expansion system. A condenser coil of the direct expansion system is cooled by glycol, and heat is rejected to the external ambient environment in a roof-mounted dry cooler. The process and devices involved are depicted in **Figure 9**.

There is also an auxiliary floor standing air conditioning (AC) unit placed in the room, as shown in **Figure 10**.

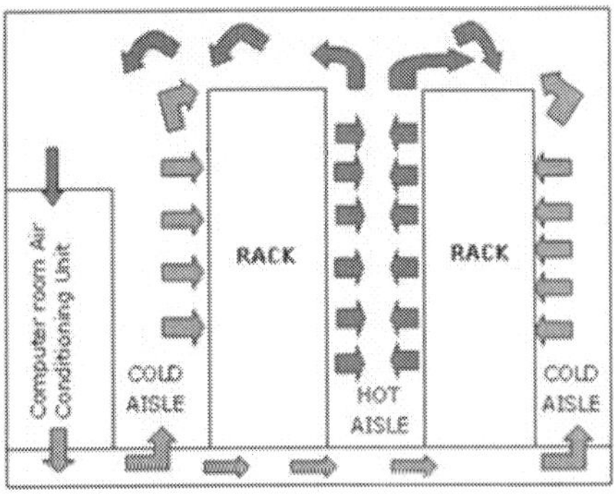

Figure 8. Schematic of hot and cold aisle arrangements without containments.

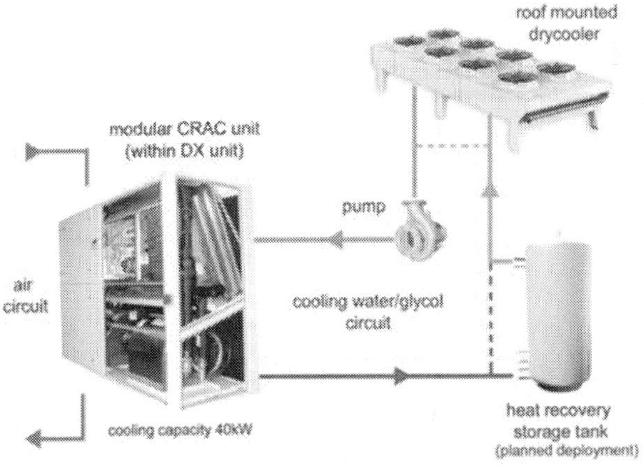

Figure 9. Main cooling system.

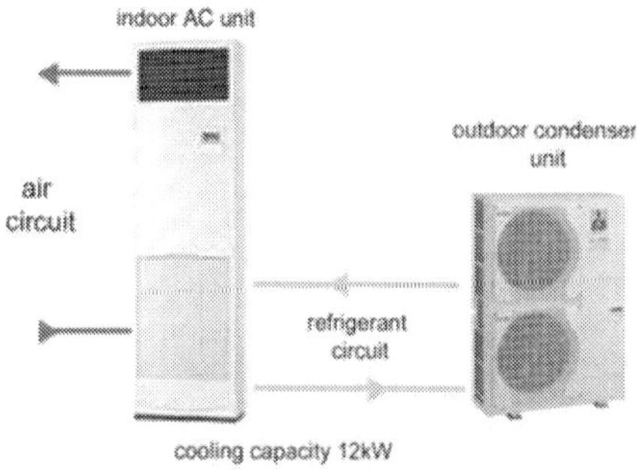

Figure 10. Auxiliary air conditioning unit.

6. Simulation-based assessment of energy management

The simulation-based evaluation of the GENiC energy management (EM) platform tests the interaction of short-term (S-T) actuation and long-term (L-T) decision-making on the virtual C130 data centre test-bed that replicates the physical processes occurring in the real data centre facility. This interaction and the components involved are shown in **Figure 11**.

A key component in all evaluations reported in this paper (and shown in **Figure 10** via the arrows between components) is the Communication Middleware GC, which provides the

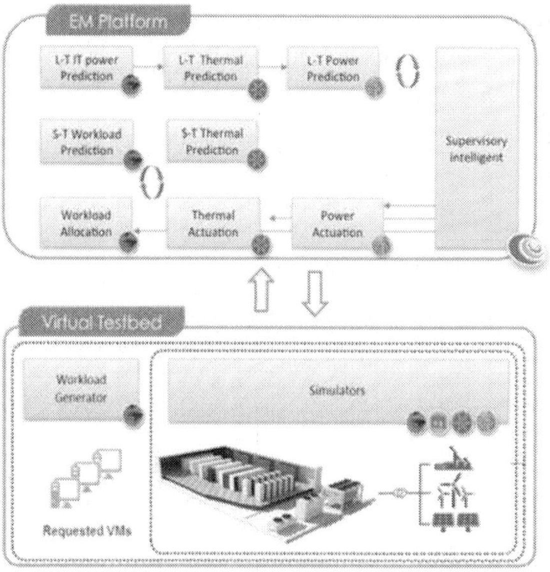

Figure 11. Interaction between EM platform GENiC components and virtual DC test-bed.

http://dx.doi.org/10.5772/65988

glue between all the different GENiC components and enables message exchange between components via the RabbitMQ broker (see above). The details of which components are relevant to a particular evaluation are discussed in the following.

6.1. Boundary conditions for the simulation-based assessment

All use cases are tested based on identical boundary conditions so that the different operating strategies can be compared to each other. The following external factors are considered as boundary conditions:

- **Requested VMs** are related to the type of services and end-user behaviour.
- **Electrical Grid info** is related to electricity market and the ratio of RES (CO_2 emission factor) in the grid.
- **Weather** conditions are specific to the DC location.
- **DC Operator Strategy** represents the baseline control strategy that establishes the reference baseline to assess the energy management saving potential.

6.2. Workload management GCG

The evaluation of the Workload Allocation Optimisation GC algorithms used within the GENiC prototype implementation was evaluated under the following scenarios (experiments):

- Workload Allocation—VM migration limits
- Workload Allocation—Thermal preferences

The experiment with VM migration limits refers to the evaluation of Workload Allocation Optimisation GC with different values for the maximum number of VM migrations allowed per time period. The evaluation with thermal preferences refers to the testing of Workload Allocation Optimisation GC considering a static thermal server preference when performing server consolidation. This experiment represents a thermal-aware workload allocation strategy [21]. The workload allocation experiment assesses the performance of the Workload Allocation Optimisation GC when it considers thermal actuation preferences. For the simulation-based evaluation, a static thermal preference matrix for each of the servers was developed based on Supply Heat Index (SHI) analysis [22] of the C130 data centre white space from the baseline inputs.

These scenarios were compared against each other and against a baseline allocation strategy. This comparison is assessed based on (i) the thermal behaviour in the white space (e.g. temperature distribution, hot spots) and (ii) energy consumption

6.2.1. GENiC components involved and testing process

The GENiC components involved in this particular workload management evaluation example are a subset of those that form the overall Workload Management GCG. This particular subset was chosen here to demonstrate the feasibility of the approach and demonstrate the overall system in operation. The experiments for this evaluation follow these steps:

1. The Simulators GC publishes the virtual time that synchronises the actions of the components involved in the experiment.

2. The Workload Generator GC publishes the VMs' resource utilization monitoring data for the current time step.

3. The Workload Allocation Optimisation GC optimizes the allocation strategy for the given arrangement in the virtual C130 DC.

4. The Workload Allocation Optimisation GC is able to consider thermal priority for each thermal box (where each thermal box represents one third of a rack). Static thermal priority is used to test a thermal awareness-based workload allocation strategy.

5. The Server Configuration component translates VM allocation to power consumption per box (one third of a rack).

The Simulators GC captures all the data relevant to this process for analysis and post-processing. The focus of this evaluation is to analyse the influence of workload allocation strategies on the temperature distribution of the white space as well as on the total DC energy consumption.

6.3. Thermal management GCG

Further experiments target the evaluation of the Thermal Management GCG algorithms with optimal thermal actuation. In this scenario, the GENiC prototype implementation is evaluated against a baseline operation strategy. This comparison is assessed based on data centre energy consumption and white space temperature distribution.

6.3.1. GENiC components involved and testing process

The GCs involved in this thermal management evaluation are a subset of those that form the Thermal Management GGCs. The subset chosen aligns with the requirements of the particular data centre demonstration site, and other, larger data centre configurations may use a broader spectrum of functionality. The experiments for the thermal management evaluation follow these steps:

1. Virtual synchronization time and current white space temperatures are published for the given time step.

2. The short-term (S-T) thermal prediction component predicts the thermal state of the white space for the next hour. This prediction supports the decision-making process that takes place in the Thermal Actuation GC.

3. Optimal temperature set points for the CRAC and AC units for the next time step are sent back to the HVAC systems model, which is part of the Simulators GC.

The Simulators GC captures all the data relevant to this process for analysis and post-processing. The focus of this evaluation is to analyse the influence of S-T prediction and thermal actuation strategies developed in the project on the temperature distribution of the white space as well as on the total DC energy consumption.

http://dx.doi.org/10.5772/65988

6.4. Power management GCG

In order to evaluate the power management aspects of the GENiC prototype platform, experiments were executed to evaluate the Power Management GCG algorithms under the following scenarios: (i) Power Actuation Logic, and (ii) Power Actuation Logic + SI static constraints. These scenarios are compared against each other and against the baseline operation. This comparison is assessed based on energy demand versus supply (broken down per source).

6.4.1. GENiC components involved and testing process

The GCs involved in this power management evaluation are a subset of those that form the Power Management GGCs and are selected to reflect the specific situation prevalent in the demonstration site. Elements of the power systems micro-grid available to the project, including a battery bank and an Organic Rankine Cycle (ORC), were modelled and included in this evaluation. The experiments for the Power Management evaluation follow these steps:

1. The Simulators GC generates the virtual time stamp and the current status of power metering for all equipment at the demand-side (DC) and at the supply-side (on-site RES).
2. The Power Actuation GC generates optimal set points for the batteries and the ORC plant for the next time step.
3. The Power Actuation GC receives a power policy (24 h profile) from the Supervisory Intelligence GC. A static SI constraint was used for the testing.

The Simulators GC captures all the data relevant to this process for analysis and post-processing. The focus of these experiments is it to analyse the power actuation operation strategies to satisfy the total DC demand. The power actuation real-time adjustments are defined so as to assure the renewable energy supply contribution. This is achieved through balancing the lack or excess of weather-dependent generation by using a controllable unit characterized with "unlimited" energy (kWh) capacity, which in this case is the ORC. The ORC has an unlimited energy capacity if the biomass storage is continuously refilled. It has to be understood that electrical batteries are characterised by limited energy capacity (here around 10 kWh) and limitations for the operation according to the definition of FSoC (fractional state of charge: between 0 and 1) upper and lower limits. According to the difference between weather-dependent renewable energy output prediction and real production, the ORC generation is adjusted taking into account the upper and lower power available referred to the maximum and minimum generation capacity of the ORC (here 4 kW minimum and 7 kW maximum).

7. Evaluation results

The simulation-based evaluation considers first results from the workload management experiments. The experimental set-up involved allocating workload over a 48-h period in a data centre using real VM resource utilization traces. Each VM was initially assigned to

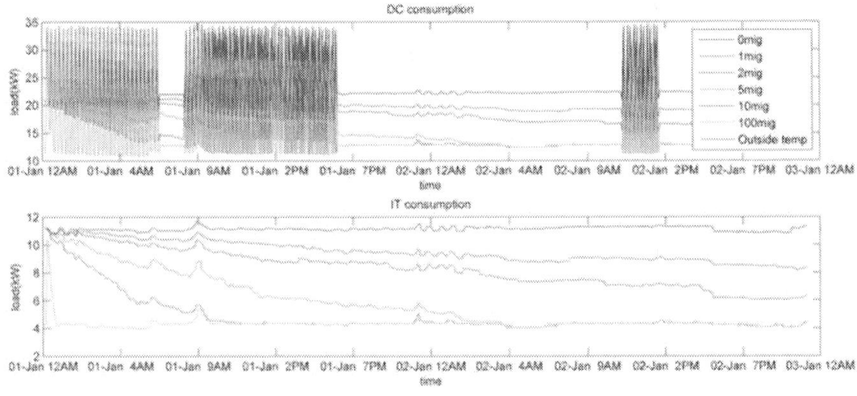

Figure 12. Power consumption with different migration limits over 48-h horizon.

a particular server as per the real traces without the Workload Allocation Optimisation GC controlling the initial assignment. The only influence on power consumption was through VM migrations and server consolidation.

7.1. Workload allocation—VM migration limits

The first experiments evaluated the impact of the migration limit on the workload allocation (without thermal priorities for servers). This baseline is a migration limit of 0, that is, each VM was run on the server it was initially assigned to. Following from there, a series of experiments were executed to evaluate various migration limits (from 1 to 100) as shown in **Figure 12**.

As expected, increasing the migration limit resulted in a considerable reduction of power consumption (see **Figure 12**). The largest migration limit tested (100 migrations per 10 min time period) required just a few time periods to achieve a reduction from approximately 11 kW to just over 4 kW. Indeed, the average hourly energy consumption of the IT equipment was 6.71 kWh less with a migration limit of 100 than with the baseline. The figure for IT power consumption (see **Figure 12**) further illustrates that all positive migration limits tended to this equilibrium state, with a migration limit of 10 reaching the 4 kW mark in less than 9 h and the limit of 5 requiring approximately 24 h. Once reached, the variations in power consumption between the migration limits were minor. This means that if the workload allocator had controlled the initial assignment of VMs to servers, then a migration limit of 10 or even 5 would have been sufficient to achieve similar savings as with a limit of 100.

7.2. Workload allocation—thermal preferences

The experiments described in the following were performed under identical settings to those previously discussed with the exception that each server had an associated thermal preference, thereby allowing a proper ranking of servers. The thermal preference was used to rank the servers for consolidation.

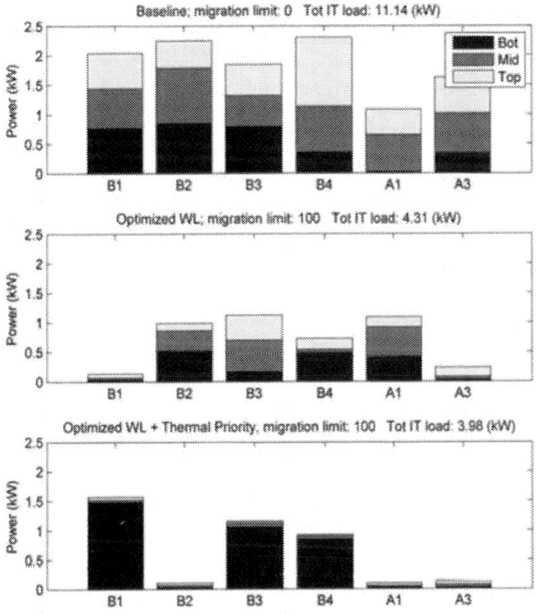

Figure 13. Workload distribution per third of rack.

In addition to the baseline described in the previous section, experiments were executed to assess power consumption with and without thermal preferences for migration limits of 10 and 100. The experiments showed that there is little difference in the total IT power consumption for the thermally ranked server consolidation, while HVAC energy consumption was reduced by approximately 20 kWh over the 48-h period relative to the baseline approach, and by 6.5 kWh compared to the scenario with 100 migrations and no thermal preference.

The behaviour of the scenarios with thermal preference can be better understood when analysed at the third of rack level (top, middle and bottom boxes) as shown in **Figure 13**. As can be observed, the only servers that were used by the GENiC energy management platform were those at the bottom level of three racks: B1, B3 and B4. The loads from all the other servers were migrated to servers in these locations and then servers that lost IT load were powered off, as can be seen from the power value for the scenario with thermal preference and limit of 100 migrations (bottom graph in **Figure 13**).

Finally, **Figure 14** presents the temperature distribution of the case study data centre C130 for (a) the thermal preference with 100 migrations and (b) the baseline. The baseline study indicates risks of a hot spot at the top layer of the last rack in row B. The supply air temperature is around 18°C; however, the inlet temperature of the particular box is approximately 23°C. The rise of temperature is due to infiltration of hot air from the hot aisle to the cold aisle space. The optimized workload allocation with thermal preference scenario ensures that the airflow

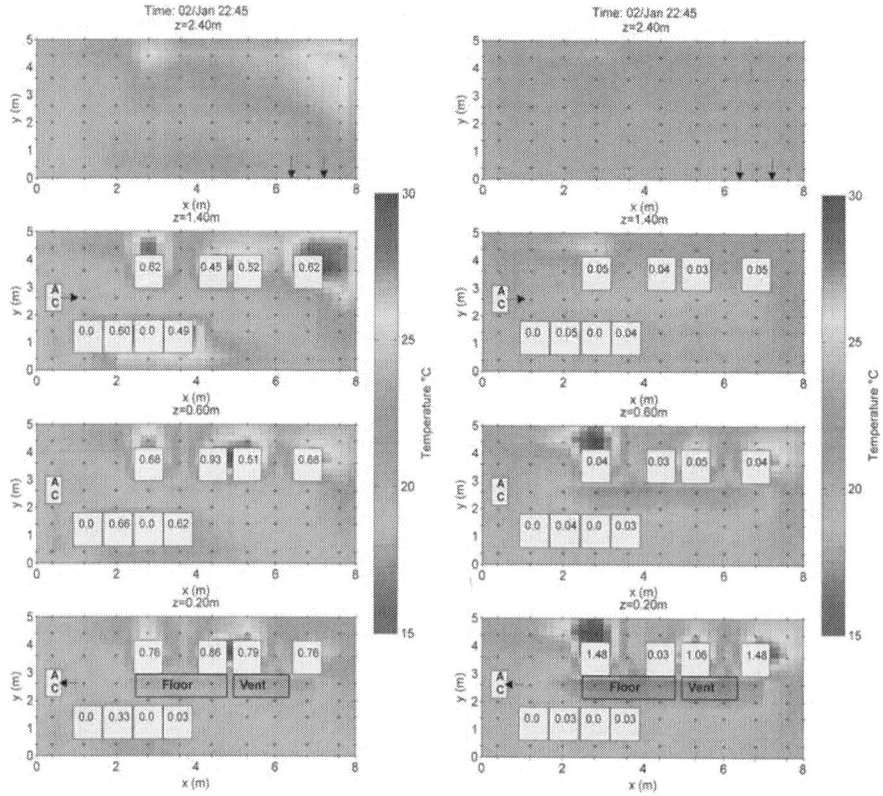

Figure 14. Temperature distribution for (a) thermal preference and (b) baseline.

will use the shortest path from the cold air supply to the heat source. The cold air is taken by preferable servers in the bottom boxes. The typical cold aisle-hot aisle distribution can be observed in this case. The inlet temperature of all active servers is approximately 18°C. This evaluation shows that the developed energy management platform can balance the temperature distribution in a data centre in such a manner as to avoid hot spots without the need for extensive structural changes to the cooling layout, for example, hot aisle containment.

8. Conclusions

In this chapter, an architecture for an integrated energy management system for data centres was presented. The architecture and prototype implementation was developed within the European Commission funded GENiC project. The proposed system combines optimisation of energy consumption by encapsulating monitoring and control of IT workload, data centre cooling, local power generation and waste heat recovery. The project conducted an initial evaluation of the platform in terms of IT workload, thermal and power management based on a simulation model of a real data centre. The initial simulation-based assessment was chosen by the project for a number of reasons. It allows evaluating the performance of management and control algorithms before deployment in the real data centre space. Secondly, the architecture of the platform is designed such that the system interacts with the simulated data centre in the same manner as it interacts with the components in a real data centre, allowing also the testing and commissioning of novel management and control concepts before deployment in target space. The specific algorithms developed in the GENiC project attempt to optimise strategies focused on workload, thermal and power management in a data centre. The optimisation occurs at different time horizons, short-term predictions are generated to support actuation decisions that are made within each of the mentioned management groups, and long-term predictions supporting decision-making at the supervisory level (coordinating management groups). The evaluation presented in this chapter focused on an initial analysis of workload and thermal management techniques. The operation strategies applied by the Workload Allocation Optimisation GC prove significant savings potential (of up to 40%) in terms of total energy consumption. This reduction is achieved through the optimization of the allocation strategy of Virtual Machines (VMs) while switching off unused servers. The performance of the Workload Allocation Optimisation GC shows a more effective utilization of the data centre with the same number of processed IT jobs. The GENiC project will replace the simulation environment by a real physical data centre for the final evaluation and demonstration of the developed management algorithms and strategies in a real-world setting.

Acknowledgements

The authors acknowledge the European Commission's 7th Framework Programme in part funding the work reported here under Grant No. 608826.

Author details

Dirk Pesch[1]*, Susan Rea[1], J. Ignacio Torrens[2], Vojtech Zavrel[2], J.L.M. Hensen[2], Diarmuid Grimes[3], Barry O'Sullivan[3], Thomas Scherer[4], Robert Birke[4], Lydia Chen[4], Ton Engbersen[4], Lara Lopez[5], Enric Pages[5], Deepak Mehta[6], Jacinta Townley[6] and Vassilios Tsachouridis[6]

*Address all correspondence to: dirk.pesch@cit.ie

1 Nimbus Centre, Cork Institute of Technology, Cork, Ireland

2 Building Physics and Services, TU Eindhoven, Eindhoven, The Netherlands

3 Insight Centre for Data Analytics, University College Cork, Cork, Ireland

4 IBM Research – Zurich, Rüschlikon, Switzerland

5 ATOS Spain SA, Madrid, Spain

6 United Technologies Research Centre Ireland, Cork, Ireland

References

[1] The Independent, "Global warming: Data centres to consume three times as much energy in next decade", http://www.independent.co.uk/environment/global-warming-data-centres-to-consume-three-times-as-much-energy-in-next-decade-experts-warn-a6830086.html

[2] Greenpeace, "How Dirty is Your Data? A Look at the Energy Choices that Power Cloud Computing. Greenpeace Report", http://www.greenpeace.org/international/en/publications/reports/How-dirty-is-your-data/, May 2011

[3] Gao, P. X., Curtis, A. R., Wong, B., Keshav S. 2012. It's Not Easy Being Green. *Proceedings of ACM SIGCOMM*, Helsinki, Finland, pp. 211–222.

[4] Qurush, A. 2010. Power-demand routing in massive geo-distributed systems, Ph.D. thesis, MIT.

[5] NRDC Issue Briefing, "America's Data Centers Are Wasting Huge Amounts of Energy", IB:14-08-a, August 2014, https://www.nrdc.org/sites/default/files/data-center-efficiency-assessment-IB.tif

[6] Uptime Institute, 2011. http://uptimeinstitute.com

[7] http://blogs.wsj.com/cio/2016/01/25/facebooks-irish-data-center-to-use-open-compute-hardware-renewable-energy

[8] https://www.google.com/about/datacenters/renewable

[9] Deng, W., Liu, F., Jin, H., Li, B., Li, D. 2014. Harnessing renewable energy in cloud data centers: opportunities and challenges. *IEEE Network*, vol. 28, no. 1.

[10] Cioara, T., Anghel, I., Antal, M., Crisan, S., Salomie, I. 2015. Data center optimization methodology to maximize the usage of locally produced renewable energy. *Sustainable Internet and ICT for Sustainability (SustainIT)*. Madrid, Spain, April 2015.

[11] Das, R., Yarlanki, S., Hamann, H., Kephart, J. O., Lopez, V. 2011. A unified approach to coordinated energy-management in data centers. *Proceedings of the 7th International Conference on Network and Services Management* (CNSM '11), pp. 504–508.

[12] Jiang, T., Yu, L., Cao, Y. 2015. *Energy Management of Internet Data Centers in Smart Grid.* Springer Verlag, Berlin.

[13] Pesch. D., et al. 2015. The GENiC Architecture for Integrated Data Centre Energy Management. *in 1st Intl. Workshop on Sustainable Data Centres and Cloud Computing (in conjunction with IEEE UCC 2015).* Cyprus, December.

[14] GENIC. 2015. GENiC public deliverable D1.4 - Refined GENiC Architecture. http://www.projectgenic.eu/

[15] Internet Engineering Task Force (IETF), "RFC7159: The JavaScript Object Notation (JSON) Data Interchange Format," March 2014, https://tools.ietf.org/html/rfc7159

[16] RabbitMQ. 2016. http://www.rabbitmq.com/

[17] A. I. Aravanis et al. 2015. "Metrics for Assessing Flexibility and Sustainability of Next Generation Data Centers," *2015 IEEE Globecom Workshops (GC Wkshps)*, San Diego, CA, pp. 1–6.

[18] J. Ignacio Torrens et al. 2016. "Integrated Energy Efficient Data Centre Management for Green Cloud Computing - The FP7 GENiC Project Experience", in 6th International Conference on Cloud Computing and Services Science (CLOSER 2016), Rome, Italy, April.

[19] Zavrel, V., Torrens Galdiz, J.I., Bynum, J.D. & Hensen, J.L.M. 2015. Model Development for Simulation Based Global Optimization of Energy Efficient Data Centres. Building Simulation 2015: 14th International Conference of IBPSA, December 7–9, 2015, Hyderabad, India (pp. 1079–1086). IBPSA

[20] Tang, Q., Gupta, S.K.S., Varsamopoulos, G. 2007. Thermal-Aware Task Scheduling for Data Centers through Minimizing Heat Recirculation. *In Proceeding of IEEE International Conference on Cluster Computing (ICCC 2007)*. Austin, TX, USA, September 2007.

[21] Sharma, R.K., Bash, C.E., Patel, C. 2002. Dimensionless Parameters for Evaluation of Thermal Design and Performance of Large-Scale Data Centers. *In 8th AIAA/ASME Joint Thermophysics and Heat Transfer Conference*. St. Louis, MO, USA, June 2002

[22] ASHRAE. 2011. TC 9.9 Thermal Guidelines for Data Processing Environments – Expanded Data Center Classes and Usage Guidance. *Data Processing*: 1–45.

Chapter 9

Thermoelectrics, Photovoltaics and Thermal Photovoltaics for Powering ICT Devices and Systems

Lourdes Ferre Llin and Douglas J. Paul

Additional information is available at the end of the chapter

http://dx.doi.org/10.5772/65983

Abstract

The conversion of heat into electricity through the thermoelectric effect and light into electricity through photovoltaic solar cells both allow useful amounts of power for a range of ICT systems from a few milli-Watts (mW) for autonomous sensors up to kilo-Watts (kW) for complete ICT computing or entertainment systems. Photovoltaics at the large scale can also be used to produce MW power stations suitable for the sustainable powering of high-performance computing (HPC) and dataservers for cloud computing. This chapter provides a background to the physics of operation of both types of sustainable energy sources along with the fundamental limits of both technologies. The present performance is presented along with promising research directions to allow for a comparison of the useful power along with the limits for deployment of each approach to power ICT devices and systems. Finally, the developing field of thermal photovoltaics is reviewed, where the overall thermodynamic conversion efficiency of turning light into electricity and useful heat can be increased through the addition of thermoelectrics or heat transfer modules to a photovoltaic cell.

Keywords: thermoelectrics, photovoltaics, energy harvesting

1. From heat to electricity: the thermoelectric effect

The increasing demand for energy has generated large amounts of carbon emissions from fossil fuels which has lead to climate change on the planet. This has made it necessary to identify new strategies to improve energy use and generation [1] which reduce the carbon dioxide emissions. Energy harvesting has become a significant field to take advantage of any free energy that is available from the environment or is waste from a system in order to recover that energy and use it for a wide range of applications. The environmental discussion of energy harvesting does not consist solely in replacing large-scale power stations and reducing their significant pollution when using coal or gas, but it also considers the use of powering smaller scale electronic devices. The idea of energy harvesting is by powering a lot of small devices, the number of large power stations can be reduced thereby reducing carbon emissions. There are other benefits too: the major one is never having to worry about recharging batteries as the harvester is constantly harvesting energy.

Thermoelectric devices are able to deliver electricity to a load using heat as a power source or to produce heating or cooling in the presence of an electrical current. The Seebeck effect converts thermal energy into electrical energy, which makes this technology suitable for harvesting energy.

The Seebeck effect was first reported by Thomas Johann Seebeck in 1821, when he observed that when two electrical conductors were brought together, and the junction between them was heated up, a small voltage reading was sensed. This effect was proportional to a constant for any material called the Seebeck coefficient (α) which was defined as the ratio between the voltage sensed (ΔV) and the existent gradient of temperature (ΔT), as define in Eq. (1)

$$\alpha = \frac{\Delta V}{\Delta T}. \tag{1}$$

Thirteen years later, in 1834, Jean Charles Athanase Peltier discovered that when an electrical current was driven through a thermocouple, a small heating or cooling was produced depending of the direction of this current. This effect, named as the Peltier effect (Π), was defined as the ratio between the heating or cooling rate at each junction (Q) and the current passing through it (I), as reported in Eq. (2)

$$\Pi = \frac{Q}{I}. \tag{2}$$

In 1855, William Thomson (Lord Kelvin) recognised the relation between the two effects explained above. These Kelvin relations demonstrated the reversible heating or cooling when there was an electrical current flowing in addition to a gradient of temperature. The relation between the Seebeck and the Peltier effect was given by

$$\Pi = \alpha T. \tag{3}$$

The Thomson effect (τ) was defined as the rate of heating or cooling per unit length through a junction, where there existed a unit current and a unit gradient of temperature. This effect was also related to the Seebeck effect by

$$\tau = T\frac{d\alpha}{dT}. \tag{4}$$

In 1911, Edmund Altenkirch derived the efficiency of the thermoelectric generation process, stating the different qualities required to define a good thermoelectric material, which will be described later in this chapter. Interest in exploiting thermoelectric phenomena for power generation began during the late nineteenth and twentieth century, and in the 1950s, the study of semiconductor materials became very interesting for the construction of thermoelectric generators, as well as practical Peltier coolers.

As the Seebeck effect is responsible for power generation, a more detailed explanation of it is given in the following section, as well as other parameters which define the efficiency of a thermoelectric system. Following this definition, a review of the different materials and approaches used during the past and present years to achieve improvements in the Seebeck coefficient is reported.

1.1. Thermoelectric power generation

Let us consider a pair of semiconductor legs (p-type and n-type) connected electrically in series and thermally in parallel. If one side of the pair of legs is heated up and the other side is kept at a reference temperature, the ΔT between the two sides produces excess carriers which may diffuse from the hot to the cold side. This diffusion of carriers sets the Seebeck voltage which will deliver a current (I) when the circuit is closed with a load, as shown in **Figure 1**.

The efficiency of the system (η) is therefore given by the ratio of the output power, P to the rate of the heat that is drawn from the source, $\eta = \frac{P}{Q}$. The current flowing through the circuit is given by

$$I = \frac{(\alpha_p - \alpha_n)(T_1 - T_2)}{R_L + R_p + R_n}, \quad (5)$$

where R_p and R_n are the resistances of each semiconductor material (p-type and n-type), R_L is the resistance of the load and α_p and α_n are the Seebeck coefficients of each leg, respectively [2]. The power delivered to the load resistor is given by Eq. (6) [2]

$$P = \left(\frac{(\alpha_p - \alpha_n)(T_1 - T_2)}{R_L + R_p + R_n}\right)^2 R_L. \quad (6)$$

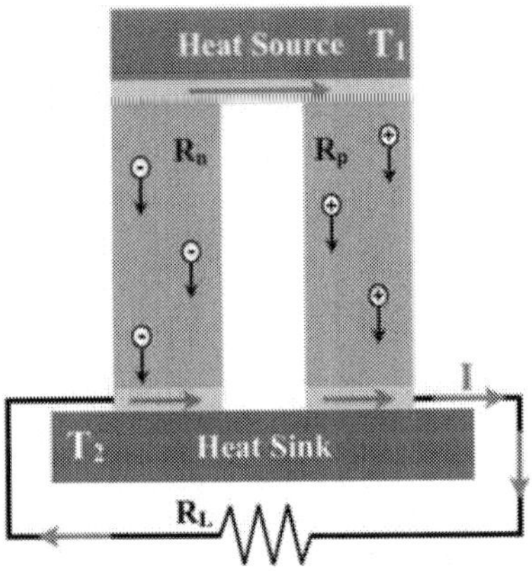

Figure 1. A schematic diagram of a module formed by a pair of thermoelectric legs connected electrically in series and thermally in parallel. The circuit has been closed, connecting a resistor across the module.

On the other hand, the heat that is drawn from the source is defined by

$$Q = (\alpha_p - \alpha_n)IT_1 + (\kappa_p + \kappa_n)(T_1 - T_2), \tag{7}$$

where κ_p and κ_n are the thermal conductances of the two legs [2].

The efficiency reaches its maximum when [2]:

$$\frac{R_L}{R_n + R_p} = \sqrt{1 + ZT} \quad \text{where} \quad ZT = \frac{\alpha^2 \sigma}{\kappa} T, \tag{8}$$

where $\sigma = \sigma_n + \sigma_p$ (S/m) is the electrical conductivity, $\alpha = \alpha_p - \alpha_n$ (μ V/K) is the Seebeck coefficient and $\kappa = \kappa_p + \kappa_n$ (W/mK) is the thermal conductivity of the material.

Using Eq. (8) in Eqs. (6) and (7), the efficiency can be defined by the following expression [2]:

$$\eta = \frac{T_1 - T_2}{T_1} \frac{\sqrt{1 + ZT} - 1}{\sqrt{1 + ZT} + \frac{T_2}{T_1}}. \tag{9}$$

From the efficiency, it is shown that if ZT is much larger than unity, the model approaches the Carnot efficiency given by $(T_1 - T_2)/T_1$. **Figure 2** shows the efficiency given for different values of ZT, where the system approaches the Carnot efficiency each time the value of ZT becomes

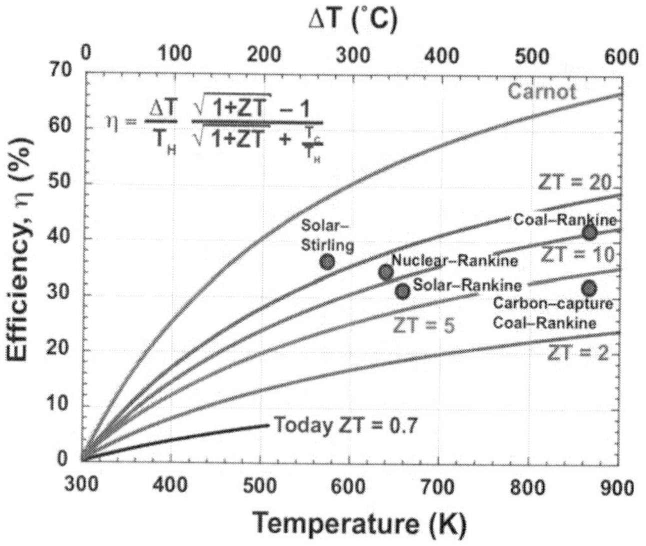

Figure 2. A plot showing the maximum thermoelectric efficiency for different ZT values. These values have been compared to the Carnot efficiency, also plotted in the figure. Also presented are the efficiencies of other technologies as a comparison.

larger. Therefore, ZT is known as the figure of merit that defines the efficiency of a thermoelectric material.

Until now, we have only considered two legs connected to a load, but a real thermoelectric generator (TEG) features several of these thermoelectric couples electrically connected in series. As the Seebeck coefficients of most materials used for thermoelectric generators are of order of 100–200 μV/K, a large number of these legs has to be connected electrically in series and thermally in parallel to generate useful voltages, > 1 V, to be able to power electronics such as ICT systems. **Figure 3** shows a diagram of a full module where several thermoelectric couples are connected electrically in series and thermally in parallel.

Getting the maximum efficiency out of a module does not mean generating the maximum power output; in fact, the power output reaches its maximum when there is electrical impedance matching between the generator and the load which occurs when $R_L=R_n + R_p$. Taking this into account, and using the relation given by Eq. (6), one gets that P_{max} is defined by

$$P_{max} = \frac{1}{2}NF\frac{A}{L}\Delta T^2\alpha^2\sigma, \quad (10)$$

where N is the number of legs, F is the fabrication factor and A and L are the area and the length of the legs, respectively (see **Figure 3**) [3]. The fabrication factor denotes the perfect system, where there are not losses of any kind, to account for contact resistances and wasted heat.

When characterising a material, apart from its efficiency, it is also important to consider separately the relation $\alpha^2\sigma$. This is the second figure of merit of a thermoelectric material and represents the output power of the system, also known as the *Power Factor.*

Current research efforts are focusing on finding new materials that can increase the efficiency (ZT) together with the power factor ($\alpha^2\sigma$) of thermoelectric devices, and that can also operate

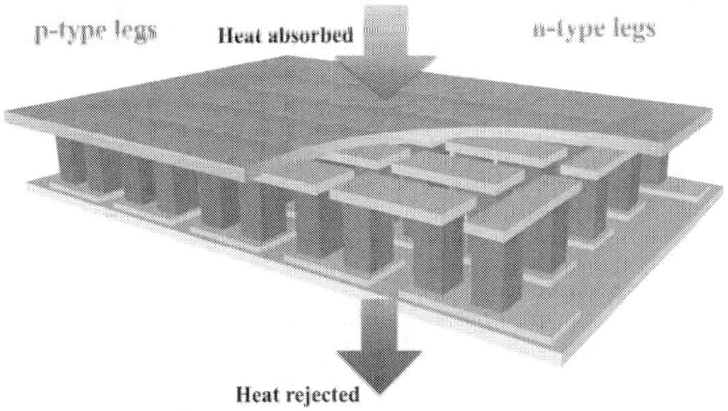

Figure 3. A schematic diagram of a thermoelectric module where the p-type and n-type legs have been bonded together, connecting them electrically in series and thermally in parallel.

over wider temperature ranges. Over the last decades, bismuth telluride and silicon-germanium materials have been extensively studied, as BiTe- and SiGe-based systems have shown the highest efficiencies at room temperatures (300 K) and high temperatures (1100 K), respectively. These platform materials are currently used in many applications, and therefore, there is significant interest in finding new materials that would be more efficient. Some research groups have been working with low-dimensional systems such as superlattices, nanowires and quantum dots, to demonstrate a new way to improve α, σ and κ; however, improving these parameters at the same time has been shown to be very challenging. A lot of effort has been undertaken to drastically decrease the thermal conductivity, which as a consequence had a good improvement of ZT, but did not necessarily mean that the power output was increased as well. In fact, it is usually found in the literature that even though the value of ZT has improved, the power factor has decreased due to the reduction in the electrical impedance matching between the generator and the load, which makes it very difficult for this material to be implemented in real applications.

Figure 4 shows a comparison of the best n-type and p-type ZT values as a function of temperature. The solid lines show the ZT values for bulk materials, where most of them present values around 1 or less than 1. The dashed lines present the values reported in the literature for low-dimensional structures where quantum effects or nanostructures can be used to optimise the ZT. The approach is always to find quantum effects or nanostructures that can scatter heat in the quantised form of phonons more efficiently than electrons, so the thermal conductivity reduces faster than the electrical conductivity. In 3D, this is nearly impossible due to the Wiedemann-Franz rule, but in low-dimensional systems, it is possible to break the linear relationship between the electrical and thermal conductivities at high carrier densities and therefore increasing ZT.

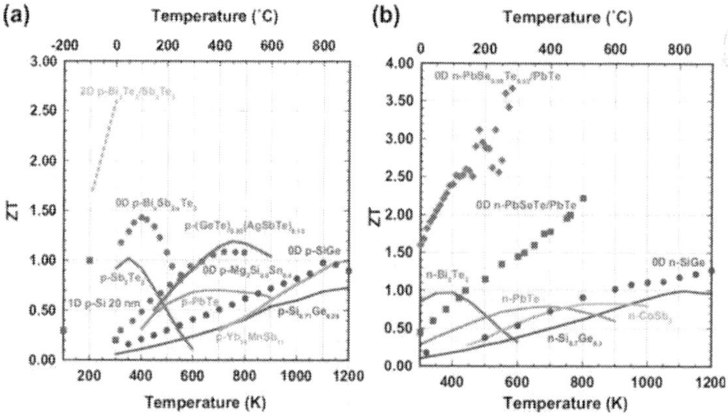

Figure 4. Left: a comparison of ZT for p-type material as a function of temperature (p-Sb_2Te_3, p-PbTe, p-$CeFe_4Sb_{12}$, p-$Yb_{14}MnSb_{11}$ [4], p-$Si_{0.71}Ge_{0.29}$ [5], 2D p-Bi_2Te_3/Sb_2Te_3 [6], 1D Si [7], 0D p-SiGe [8], p-$(GeTe)_{0.85}$ $(AgSbTe)_{0.15}$ [3], 0D p-$BixSb_{2-x}Te_3$ [9], 0D $Mg_2Si_{0.4}Sn_{0.6}$ [10]). Right: a comparison of ZT for n-type material as a function of temperature (n-Bi_2Te_3, n-PbTe, n-$CoSb_3$ [4], n-$Si_{0.7}Ge_{0.3}$ [5], 0D PbSeTe [11], 0D n-SiGe [12], 0D n-$PbSe_{0.98}Te_{0.02}$/PbTe [13]).

1.2. Applications

Thermoelectric generators are robust, do not have moving parts, do not require maintenance and can generate continuous power as long as there is a heat source. Therefore, this technology is an attractive way to recover wasted heat rejected into the environment and is normally a fit and forget technology.

Thermoelectric generators can be used over a wide range of temperatures, which makes them useful in many different systems. In the following list, there are some of the applications mentioned where thermoelectric generators are currently used or are under investigation.

1.2.1. Applications close to room temperature

- Implantable medical devices have the disadvantage of depending on batteries, with life times ranging from 5 to 10 years. These devices could be powered by using temperature differences that exist between the inner surface of the skin and the body core. A thermoelectric module generating around 70 μW in the presence of these temperature gradients could be useful in these applications [15].
- The major application for thermoelectric devices at present is as Peltier coolers (**Figure 5** left) to maintain electronic and optoelectronic components at stable temperatures. For instance, laser diodes are kept at a constant temperature by Peltier coolers to obtain a constant emission wavelength for telecoms applications. Other applications, such as DNA amplifiers for cell imaging, have a strong reaction sensitivity to temperature, and therefore, Peltier elements are used not only for temperature stability but also to enable a large range of temperatures that they can provide. Thermoelectric cooling is also used in the automotive industry to cool down the batteries of electric cars when these are charging.
- Wireless sensors are autonomous devices combining sensing, power, computation and communication into one system; smartdust has become a term to refer to these kind of sensors. Most wireless sensors systems now require between 1 and 5 mW of power to run mainly dependent on the distance for the communication, and so a cm^2 area thermoelectric

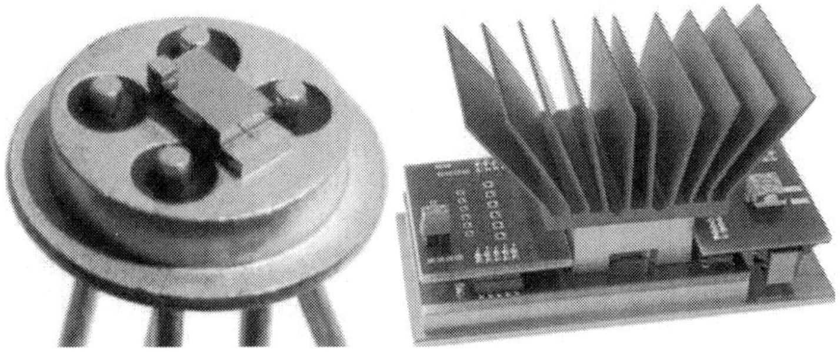

Figure 5. Left: A telecoms laser on a microfabricated Peltier cooler produced by Micropelt. Right: A thermoelectric generator produced by Micropelt showing the cold sink aimed at dissipating heat through air cooling. copyright Micropelt [14].

device requires around 50°C to provide sufficient power. As the communications consume the most power, most systems have rechargeable battery or super-capacitor storage systems and then use burst modes of communication so that information is only sent when required thereby minimising the power consumption. Therefore, it is the cost of replacing batteries (mainly labour costs) that allows thermoelectrics and other forms of energy harvesting to be cost-effective. As an example, EnOcean described how after installing 4200 energy harvesters to power light switches, occupancy sensors and daylight sensors in a new building, they had saved 40% of lighting energy costs, 20 miles in cables and 42,000 batteries (over 25 years), and as a consequence, they had reduced the amount of toxins released by batteries to the environment [1]. The right image in **Figure 5** shows a thermoelectric generator embedded onto a circuit board to collect part of the heat generated by the electronics and convert it into useful power.

1.2.2. High-temperature and industrial applications

- The largest driver for improved thermoelectrics at present is probably the car industry where European legislation to improve fuel efficiency is driving thermoelectrics research to replace the alternator. The car is a system where thermoelectrics could play a big role as 75% of the fuel ends up as waste heat and the 40% of waste heat goes down the exhaust pipe into an environment that could be used to capture this heat and convert it into electricity, as shown in (**Figure 6**) [3, 6, 16]. The temperature of the exhaust system can range from room temperature up to 750°C, so this is driving work on new thermoelectric materials to replace the best at present which is PbTe with toxic Pb that cannot be used for applications. Initial modelling has suggested that up to a 5% in fuel consumption could be achieved with suitable thermoelectrics with ZT of 1, but the key issue is getting the whole

Figure 6. A schematic diagram of how the energy from a combustion engine in a car is distributed. 25% of the energy produces motion and through the alternator generates electricity to power accessories including the electrics, the air conditioning and the hifi system. 75% of the energy from the fuel is lost mostly through friction and heat. 40% of the fuel energy disappears through the exhaust system; hence, there is interest in using thermoelectrics to harvest some of this waste energy.

thermoelectric system cheap enough for the market. Also, no thermoelectric provides ZT of 1 from room temperature to 750°C, so segmented modules and/or new materials are required. Most of the major car companies are now working heavily of thermoelectric, and it is only a mater of time before automotive systems become available.

- Due to the absence of vibration, noise or torque during operation, thermoelectric generators are suitable systems for powering space missions [3]. Space systems use radioisotope thermoelectric generators (RTGs) where a radioactive material heated by the decay, and emission of radiation is used as the hot source with a thermoelectric generator to turn the heat into electricity. These systems typically operate close to 1000°C, and because of these high temperatures, SiGe has been the main thermoelectric material used for these generators, reaching efficiencies as high as 6.6%. Therefore, this is the major reason NASA used radioisotope thermoelectric generators for the Voyager space probes which have been operating for over 34 years and have now left the solar system.

2. From light to electricity: the photovoltaic effect

Solar cells are semiconductor devices which are able to convert solar energy into electricity [17], using the photovoltaic (PV) effect which is responsible for this conversion. When a p-n junction formed by a semiconductor material is exposed to an electromagnetic radiation, such as light, photons with sufficient energy will excite electrons from the valence band to the conduction band leaving a hole behind. Due to the built-in asymmetry of the device, the excited electrons and holes will be separated away to an external circuit to create an electrical current [18]. The efficiency of a PV device depends on the light absorbed by the material and also on the electrical connections to the external circuit.

The PV effect was discovered by Edmund Becquerel in 1839, who observed that a small electrical current was generated after exposing an electrode to light which was immersed in an electrolyte solution [19]. Some years later, in 1877, William Adams and Richard Day built the first solar cell, which consisted of a selenium sample with two platinum contacts on it [20].

The efficiency of these first devices, however, was still too poor for any applications, and therefore, it was necessary to wait many decades in order to have a better understanding of semiconductor materials. In 1947, William Shockley, John Bardeen and Walter Brattain developed the first germanium transistor [21], which led to a method to fabricate p-n junctions, first in germanium and later in silicon, with better photovoltaic behaviour. In 1954, Chapin, Fuller and Pearson developed the first silicon solar cell in the Bell Telephone laboratories. This solar cell was the first photovoltaic device to convert light into electrical work with an efficiency of 6% [22]. The interest in solar cells drastically increased due to the oil crisis in the 1970s. Alternative energy sources were required, and therefore, a lot of funding was invested into solar energy development.

At first, solar cells were only used for space applications, but soon, the efficiency and cost of silicon solar cells were improved to the point which made possible their use in many terrestrial applications. Other semiconductor materials, such as GaAs, and thin film technologies were investigated, as well as new structures with multiples junctions to widen the absorption of radiation and hence improving the efficiency of these devices.

During this period, new structures were developed to improve the efficiency of these PV devices. Some of the structures that are still currently in use are the development of thinner junctions to improve the absorption of ultraviolet light and the development of textured surfaces as well as antireflective coatings to decrease the losses due to the reflection of light [23].

In the 1980s, silicon solar cells with efficiencies around 20% were first fabricated. Currently, silicon is the cheapest, most abundant and available semiconductor material with a useful PV efficiency, and therefore, it is the most used material for terrestrial applications. In 2014, 35% of the PV market used crystalline silicon and 56% of the market used poly-crystalline silicon which combined account for 91% of the overall PV market [24]. The remaining 9% of the market is from thin film technologies which includes amorphous silicon and CdTe. The development of new manufacture technologies, however, as well as the investigation of new materials that could improve the efficiency and reduces the cost are still in development.

2.1. Available solar energy

The sun emits light within a range of wavelengths where three regions can be differentiated: the ultraviolet, visible and infrared region of the electromagnetic spectrum, see **Figure 7** (left). **Figure 7** (right) shows the solar irradiance as a function of wavelength at a point outside the Earth's atmosphere (AM0, blue line), which corresponds approximately to the radiation of a black body with a temperature of 5760 K [also shown in **Figure 7** (right, black line)]. The sun (with a temperature of 5760 K) presents its strongest emission at visible wavelengths, reaching its peak inside the blue-green region.

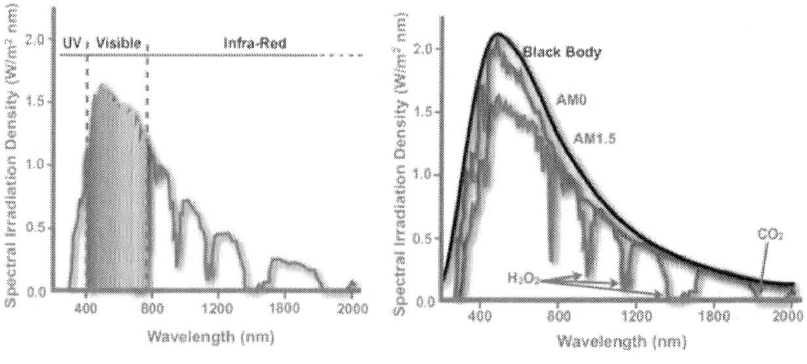

Figure 7. Left: spectral irradiation density at AM1.5 versus different wavelengths, showing the UV, visible and infrared region. Right: spectral irradiation density versus different wavelengths for AM0 and AM1.5, and for a black body with a temperature of 5760 K.

http://dx.doi.org/10.5772/65983

As solar energy travels through the atmosphere, light is absorbed by many of its elements. In fact, some of these elements can be identified by their absorption lines (Fraunhofer lines). As an example, water and CO_2 are mainly absorb in the infrared region, being responsible for some of the absorption dips in the spectrum showed in **Figure 7** (right, AM1, red line).

The attenuation caused by the light traveling through the atmosphere is defined by the 'Air Mass' index, which is defined as:

$$n_{AirMass} = \frac{\text{optical path length to sun}}{\text{optical path length if sun directly overhead}} = \frac{1}{cos(\gamma)} \tag{11}$$

where γ (see **Figure 8**) is the angle between the optical path length of sun and the normal through a horizontal plan at the point of observation.

As the spectrum of light changes depending on the day and the location, a standard reference for the spectra is defined in order to perform valid comparison of PV devices within the different research institutes and companies. For terrestrial use, the standard reference is defined for an air mass of AM1.5, which corresponds to an angle of elevation for the sun of 42° which is a latitude of 37°, and to an irradiance of 1000 Wm^{-2}.

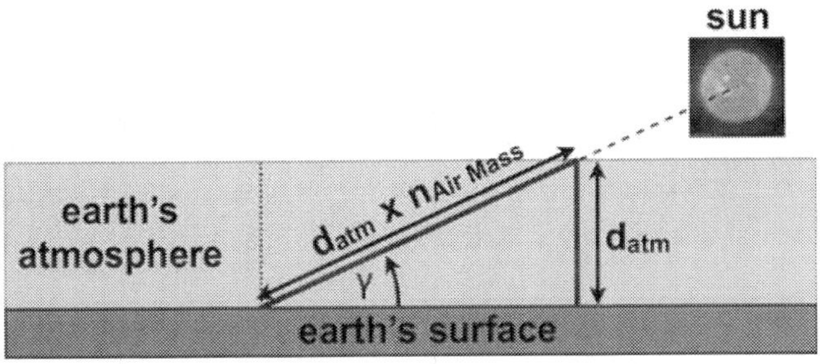

Figure 8. A schematic diagram showing the atmospheric attenuation and how the air mass index is calculated.

2.2. P-N junction

Semiconductor materials are used in order to take advantage of the electromagnetic radiation that comes from the sun to convert part of this energy into electricity (the photovoltaic effect). The physics for this effect to take place inside a material is based on the p-n junction, and the junction between two semiconductor materials doped n-type and p-type [18, 25].

In a semiconductor material, the lower energy level (valence band, E_v) is separated from the energy level where an electron can be considered free (conduction band, E_c), by an energy gap known as the band gap (E_g). When the semiconductor material is exposed to a flux of light, photons can be either reflected, absorbed or transmitted, see **Figure 9** (left). If the photons are reflected or transmitted, then they will be lost into the environment without contributing to the PV current. On the other hand, if they are absorbed by the material [**Figure 9** (right)], only

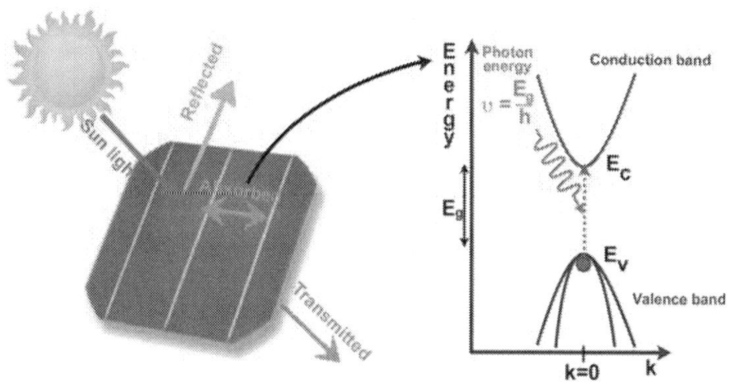

Figure 9. Left: A schematic diagram showing how the light can be either reflected, transmitted or absorbed by a solar cell. Right: Only photons which are absorbed can contribute to electron-hole generation. A photon with a greater energy than E_g can excite an electron from the E_V to E_C.

photons with a frequency υ corresponding to an energy ($h\upsilon \geq E_g$) equal or greater than the band gap will be able to excite an electron from the valence band into the conduction band, and therefore contribute to the PV current.

The concentration of electrons inside the conduction band and the concentration of holes inside the valence band can be controlled by the addition of impurities into the material. These impurities are considered either as donors (N_D), which increase the concentration of electrons inside the conduction band, or acceptors (N_A), which increase the concentration of holes inside the valence band. If the concentration of electrons is higher than the concentration of holes, then the current is dominated by the movement of electrons, and the semiconductor is considered as n-type. On the other hand, if the current is dominated by the movement of holes, then the semiconductor will be considered as p-type.

If a photon has enough energy to create an electron-hole pair, then there is a high possibility for the carriers to be separated by the electric field generated by the p-n junction, which will contribute to generate an electric current once the device is connected to a load (**Figure 10**). In reality, a p-n junction only provides a very small depth inside the depletion region for absorption so to increase the absorption depth of the PV cell, p-i-n junctions are formed with a large undoped i absorption region between the two doped regions. There is also a possibility for the excited electron to recombine with a hole inside the valence band before the electric field can drive away the carriers. This process is known as recombination, and it is a loss mechanism for solar cells as the generated electron hole pair is re-emitted as light from the cell.

The distance that a photon can travel inside the material before it is absorbed is defined as the absorption coefficient (α), Eq. (12):

$$\alpha = \frac{4\pi\kappa}{\lambda} \tag{12}$$

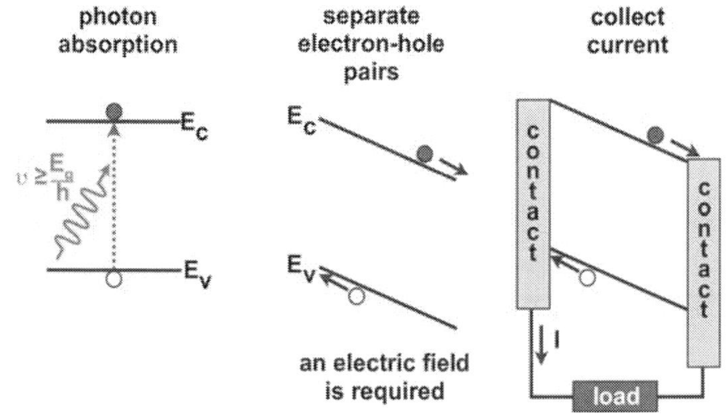

Figure 10. A schematic diagram showing a sequence of steps (from right to left) from photon absorption into the material, to electron-hole pair generation and to collection of current into an external circuit.

where κ is the extinction coefficient and λ is the wavelength. The absorption coefficient not only depends on the material, but also on the wavelength at which light is to be absorbed. For instance, if the material is too thin, photons with high wavelength will be transmitted as if the material was transparent to them. **Figure 11** shows the absorption coefficient values as well as the penetration depth ($1/\alpha$) into the material as a function of photon energy for a range of semiconductor materials typically used to make PV cells. Whilst Si is the material which dominates the PV market, it can be observed that it does not have the best absorption coefficients for PV cells. Ge, InP and GaAs have better absorption coefficients for a given photon

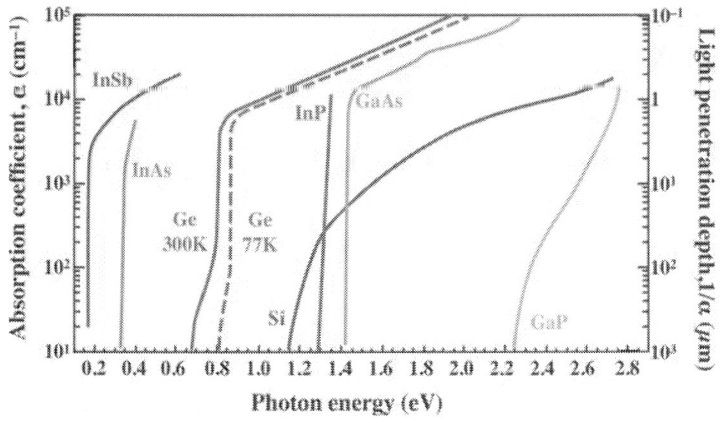

Figure 11. Absorption coefficient and penetration depth values as a function of photon energy for different semiconductors.

energy allowing less material, but Si dominates because it is one of the cheapest and most plentiful material available in the semiconductor industry.

2.3. Circuit model of p-n photovoltaic solar cell and the efficiency

A photovoltaic device is based on a semiconductor material p-n or p-i-n junction where an asymmetric junction is required in order to separate the carriers. Most of these devices behave as a diode in the dark, where the device admits more current under a forwarded voltage than at a reversed voltage. Therefore, it can be said that a solar cell behaves as a current source with a diode connected in parallel [see **Figure 12** (left)].

When the solar cell is not connected to a load (open circuit), most of the generated current flows through the diode, which is known as the dark current. The dark current is equivalent to the current that would flow through the circuit if an external voltage was applied to the cell in the dark. Eqs. (13) and (14) define the current (J_{dark}) and voltage (V_{oc}) when the device is in open circuit

$$J_{dark} = J_0(e^{qV/k_BT}-1) \quad (13)$$

$$V_{oc} = \frac{k_BT}{q}\ln(\frac{J_{sc}}{I_0}+1) \quad (14)$$

where J_0 is a constant, k_B is the Boltzmann's constant, T is the temperature and J_{sc} is the short-circuit current. As an approximation, the current-voltage characteristic is the superposition of the dark current plus the short circuit photocurrent [J_{sc} **Figure 12** (right)].

Looking at the equivalent electrical circuit represented in **Figure 12**, the total current density is given by

$$J_V = J_{sc} - J_{dark}(V), \quad (15)$$

which becomes the following expression for an ideal diode,

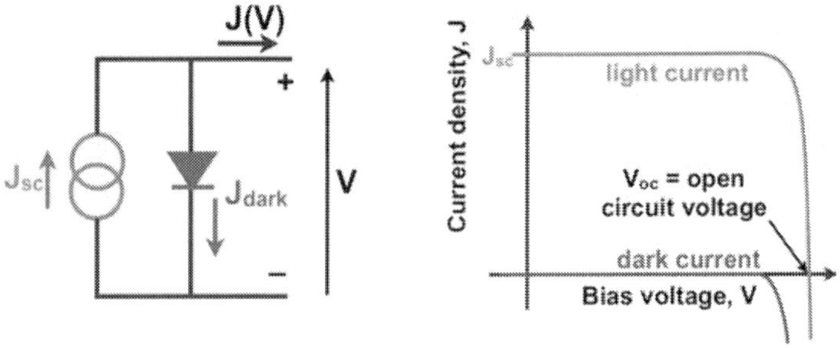

Figure 12. Circuit model for a solar cell. The devices behave as a current source with a diode connected in parallel.

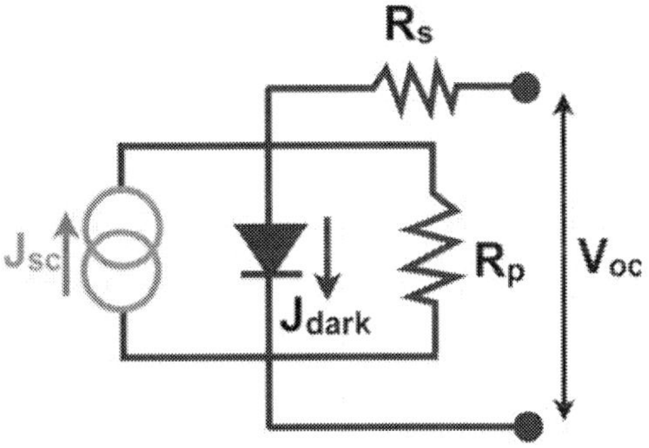

Figure 13. The circuit model for a real solar cell, where the series resistances due to the contacts and the parallel resistances due to the p-n junction have been added.

$$J_V = J_{sc} - J_0 (e^{qV/k_B T} - 1). \quad (16)$$

Until now, we have only discussed ideal PV devices, but unfortunately, real solar cells present two parasitic resistances which degrade the performance of these devices. **Figure 13** shows the equivalent electrical circuit for a real solar cell, where there is a series resistance (R_s) and a parallel resistance (R_p). R_s is related to the series resistances created by the front and back contacts between the metal and semiconductor and also due to the interconnects used for connecting the electrical circuit. R_p is related to the leakage current of the cell around the edges of the device, and it also shows the quality of the p-n junction.

The power density of a solar cell is given by

$$P = JV. \quad (17)$$

The maximum power density for a photovoltaic device occurs at a certain V_m and I_m, which does not correspond to the maximum current or voltage provided by the device, see **Figure 14**. Therefore, the fill factor (FF) is defined as the fraction between the maximum power that the solar cell can deliver to a load, and the theoretical maximum power density defined by J_{sc} and V_{oc}

$$FF = \frac{V_m J_m}{V_{oc} J_{sc}}. \quad (18)$$

The more the FF approximates to 1, the more the power density approaches to the theoretical value. The efficiency then can be defined as the maximum power that a solar cell can deliver to a load at an operating point P_m versus the incident light power density P_s

$$\eta = \frac{P_m}{P_s} = \frac{FFV_{oc}I_{sc}}{P_s}. \quad (19)$$

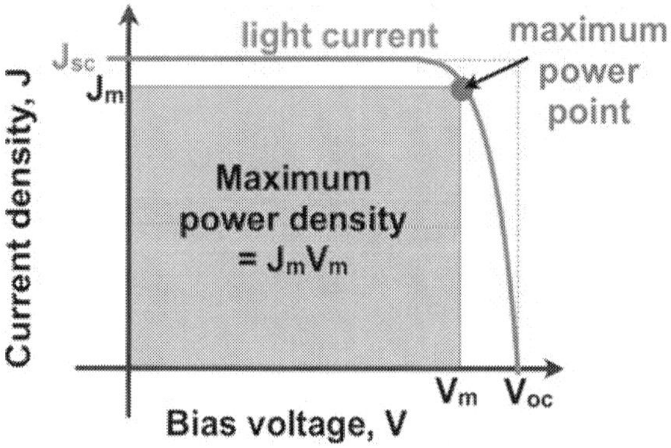

Figure 14. Schematic diagram showing the maximum power point of a solar cell, which does not match with the theoretical maximum power density.

Therefore, J_{sc}, V_{oc}, FF and η are the parameters evaluated to define the quality of a solar cell. As explained in Section2.1, solar cell is normally tested in laboratories following the AM1.5 spectrum, with an incident light power density of 1000 W m^{-2} and a temperature of 25°C.

Figure 15 shows the current density values versus the open-circuit voltage values for different semiconductor materials.

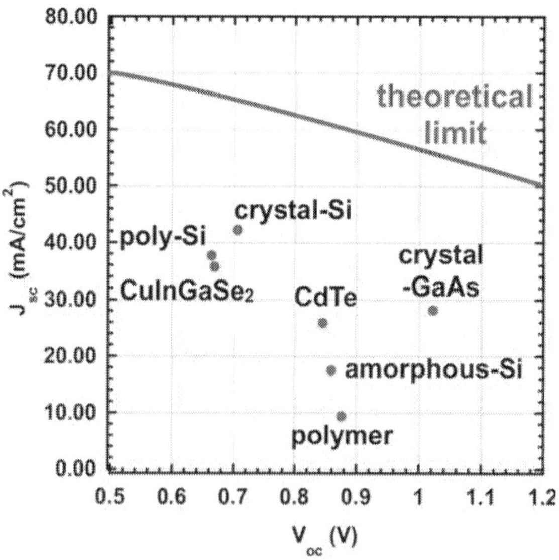

Figure 15. Graph showing the current density values as a function of V_{oc} for different semiconductor materials.

2.4. PV efficiency

The Carnot efficiency of a PV cell can be calculated by considering the PV cell as a perfect blackbody absorber. Since the sun emits a blackbody spectrum with a temperature corresponding to 5760K, it can be demonstrated that the maximum absorption occurs when the PV cell as a blackbody absorber is at 2470K. The Carnot efficiency is 85% when all the photons are absorbed, and each absorbed photon generates the maximum amount of heat available from the photon, and there is zero thermal dissipation from the absorber. Of course, no real PV cell can be operated at 2470K, and many of these approximations are not true in the real world so all efficiencies will be lower than this number.

A more practical efficiency is for a PV cell which produced from a single semiconductor material. This was first calculated by Shockley and Quessier [26] in 1961. The calculation assumes a PV cell which emits as a blackbody and has no light incident and no applied voltage. The work then considers all the radiation which can be absorbed by the semiconductor with energies above the bandgap, E_g, and assumes only band-to-band recombination. Geometrical factors are added to work out the amount of light that can be absorbed by a flat PV cell and also to account for the temperature of the sun and the temperature of the PV cell. For PV cells at 300K, the maximum efficiency for a Si cell is 30% and for a GaAs cell is 31%. The Shockley-Quessier efficiency is a maximum for a bandgap of about 1.4eV and falls off for larger and smaller bandgaps. The best Si cell is 25.6%, and the best GaAs single cell is 28% for illumination with 1 sun [27], so these numbers are quite close to the Shockley-Quessier limit.

For any material, a number of optimisation strategies are employed in all PV cell manufacture to increase the performance and get close to the Shockley-Quessier limit. If a piece of flat silicon is used as a PV cell, then the maximum efficiency is around 13%, well below the record 25.6% efficiency [27]. Key to optimising the cell is to realise that not all the light from the sun enters the cell, and even if the light does enter the PV cell, not all of the photons will always be absorbed. The first approach is to texture the surface which increases the path length of the light inside the PV material if a metal reflector is deposited onto the back surface. The left side of **Figure 16** presents the concept of textured surfaces where the sunlight has the rays reflected when absorbed. The light is then reflected along with the material from a back reflector, and if

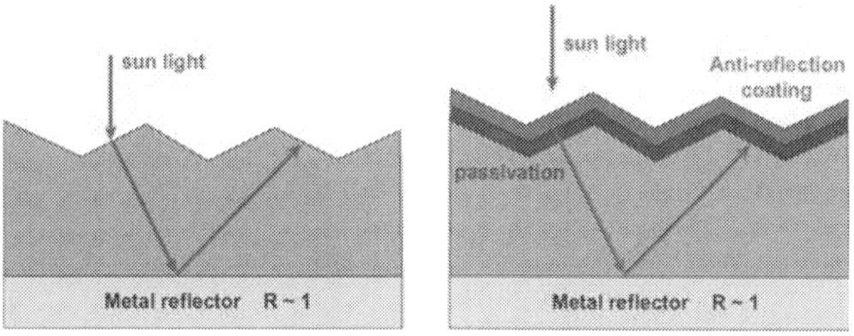

Figure 16. Schematic diagrams showing how a corrugated surface can be used to increase the path length of light in the solar cell and therefore increase the absorption of radiation. Left: a simple textured surface which increases the path length of light through refraction at the surface and a backside metal mirror. Right: the same correlation but with an additional anti-reflection coating to reduce the reflected light from the cell surface.

the correct angles are chosen for the texture, then total internal reflection can increase the path length to typically four times the thickness of the material. This is especially important for indirect bandgap materials such as silicon where the absorption coefficient (see **Figure 11**) can be quite low close to the bandgap edge.

The second key part of optimising a PV cell is surface passivation. Dangling bonds at the surface where the periodic crystal lattice is broken can trap the photo excited electron-hole pairs resulting in a reduction in the efficiency of the PV cell. Passivation is important to reduce the number of surface traps to provide the maximum efficiency. For Si PV cells, thermal silicon dioxide provides a very high-quality passivation where surface trap densities can be reduced to $\sim 10^{10}\,cm^{-2}$ whilst such a low level of traps is difficult to achieve on other semiconductor materials.

Finally, an anti-reflection coating can significantly reduce the light that is reflected from the PV cell surface (see **Figure 11** right). The reflection R from a surface is dependent on the refractive index of the cell material n_{PV} and the refractive index of air which is $n_{air} \sim 1$. The reflection from the PV surface is given by

$$R = \left(\frac{n_{PV} - n_{air}}{n_{PV} + n_{air}}\right)^2. \tag{20}$$

For Si, $n_{PV} = 3.5$ and so 30% of the light is reflected from a silicon surface leaving only 70% of the sunlight able to be absorbed by the PV cell. The way to reduce the amount of light which is reflected is to add an anti-reflection coating. The reflection R vanishes if a coating is deposited with an intermediate refractive index given by $n_{ARC} = \sqrt{n_{air}.n_{PV}}$ with a thickness of a quarter of the wavelength, $\lambda/4$. Such coatings can easily allow > 95% of the light to enter the PV cell at λ, but it is difficult to produce an anti-reflective coating that has high absorption over the whole of the spectrum of the sun.

The Shockley-Quessier limit is only for a single semiconductor material at 1 sun illumination. This efficiency can be increased if concentrators are used to increase the intensity of the light that impinges onto the PV cell surface. Concentrators can increase the maximum efficiency from roughly 31% at 1 sun to ~40% at 1000 suns illumination. A number of concentrator systems does operate with efficiencies well above the Shockley-Quessier limit of 31% for 1 sun [27].

The second approach to get efficiencies above the Shockley-Quessier limit is to use more than 1 semiconductor material with different bandgaps. Stacked systems with spectrally selective mirrors are one way to improve the spectral absorption and overall efficiency but such systems are much more expensive than multi-junction PV cells. Here, 1 or more semiconductors are epitaxially grown onto the one substrate to provide more than one bandgap for absorption. The widest bandgap material must be at the surface, and the record for a 5 junction solar cell is now 38.8% for 1 sun illumination and 46% with a 300 concentrator [27]. Whilst these recent results are now quite impressive, the multi-junction technology is predominantly with III-V materials and is therefore very expensive. It is not yet at the level to help to significantly reduce the cost per Watt which is key for applications and the market.

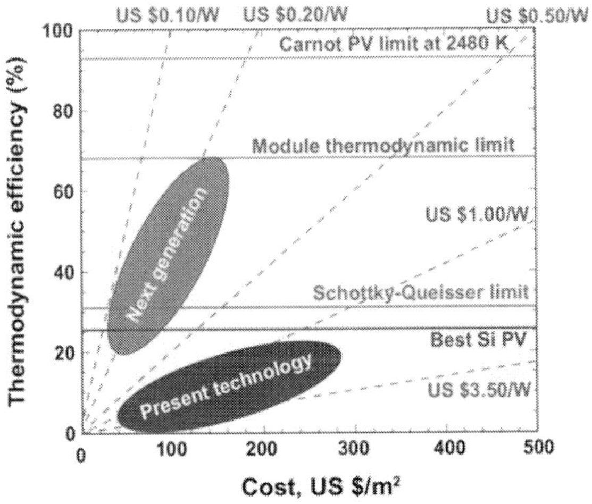

Figure 17. Graph showing the current density values as a function of V_{oc} for different semiconductor materials.

Figure 17 shows the cost of PV technology per unit area versus the efficiency. Also marked on the figure is the cost per Watt along with some of the key efficiencies for present Si PV, the Shockley-Quesser limit, the Carnot limit and the realistic module limit for multi-junction cells. Present crystalline Si, amorphous Si and other thin film technologies have efficiencies from 10 to 22% available on the market. For next generation technologies to have any major impact, they must manage to reduce the cost per Watt and also the capital cost of the PV technology per unit area. No technology has yet managed to achieve the low costs and the high efficiencies, so this is still a research area ripe for new technologies.

2.5. Applications

PV solar modules have a wide range of applications, many of which are well outside of ICT systems but also they can be used for powering a wide range of ICT systems.

- Large-scale solar farms for the powering of data centres and HPC clusters: 10MW solar farms can certainly be built in appropriate environments, and many data centres already use the technology to reduce the carbon emissions. At present, the generation levels and poor large-scale electricity storage prevent 24/7 operation of the data centres from the PV renewable technology, and diesel generators or grid electricity are required for night operation when no solar energy is generating electricity.
- Small-scale ICT systems for indoor use such as calculators and digital watches: These applications have been around since the first products in the 1970s and can be enabled from the low-power requirements of the electronics.

- Rechargeable PV systems for mobile phones and larger ICT systems of the Watt and a few 10s Watt level: These are PV systems of 10–100 cm^2 which require direct sunlight to be able to generate the Whr required for ICT systems of the W to 10s Watt level. PV is really the only renewable or energy harvesting technology that can achieve power levels at the W level whilst thermoelectrics and vibrational systems have to be enormous and only really in industrial environments to be able to achieve power levels close to these values. Such PV systems can also recharge tablets and small laptops, but the charging time may be significant for the larger systems.
- Autonomous sensors both indoor but more frequently outdoor use PV cells with super capacitors or rechargeable batteries for continuous operation. Many road and rail warning and information signs as well as public information notices now use PV cells to provide a fit and forget technology to power the systems. Some building autonomous sensors for fire, smoke and temperature are now being sold with PV systems.
- Both industrial and domestic properties use PV to generate electricity. Whilst the amounts seldom are large enough to power the whole building, PV can generate sufficient energy to power a range of items in any building and in doing so be useful to reduce the load to the national grid and electricity generated from fossil fuels. In many cases, it is sensible to power ICT devices as these require DC power, and the DC generation from the PV does not require to be converted to ac and then back to DC with the combined losses in both conversion processes. In many countries, feed-in tariffs have been used to subsidise the large capital costs to enable countries to significantly increase their renewable energy generation capacity.

3. Thermal photovoltaics (TPV)

It is clear that neither thermoelectrics nor photovoltaics are operated anywhere close to the Carnot limit for thermodynamic conversion (see **Figures 2**and **17**). Therefore, both systems have significant losses which result in heat. A secondary issue is that the thermal to electrical conversion efficiency of photovoltaics decreases as the temperature is increased. For example, a crystalline Si PV cell reduces in efficiency by about 0.4% per 1°C rise [28]. Therefore, there has been an interest in combining photovoltaics and thermoelectrics together in the field of thermal photovoltaics (TPV). The whole of the TPV field is much larger than just PV combined with thermoelectrics and includes thermal heating for combined heat and power systems especially for integrated modules into buildings where the heat is used either to produce hot water for a property or to heat the building [28]. A number of PVT technology is optimised to maximise the heating element which is not appropriate for ICT systems. Here, we will restrict the use to energy harvesting scenarios which are optimised and appropriate for providing electricity to ICT devices and systems. With this proviso, the range of PVT technologies which are useful reduces dramatically as a key part must be the production of electricity.

As has been discussed with PV, concentrators allow significant improvements in the efficiency and output power. Whilst improvements are obtained with single junction technology, the largest power outputs are obtained with concentrator-based multi-junction PV. Such technologies are not appropriate for miniature portable PV systems but tend to be more appropriate for large-scale solar farms, many of which are used with HPC or data centres. Whilst the concentration increases the conversion efficiency per unit area and increases the output power, the disadvantage is the significant increase in operating temperature of the PV which both reduces efficiency and can also lead to failures through repeated thermal expansion and contraction. The thermal cycling failures can be more significant where ice formation at night can occur since the volume of ice is significantly larger than an equivalent mass of water leading to significant mechanical damage if the ice grows in particular ways around PV modules. Therefore, one form of TPV is to use a thermal store with a heat pump such as a thermoelectric to reduce the temperature of the PV section thereby increasing the efficiency but then storing this heat energy during the day. Heat is pumping back to the PV at night to reduce the probability of ice formation and failures from thermal contraction.

The most common form of a PVT generator at present is a PV cell with a fluid-based system to conduct the heat away from the cell to use for heating buildings or hot water systems. The best of these systems have an overall efficiency approaching 70% with about 20% electrical efficiency and over 50% thermal efficiency. The priority for ICT applications similar to most of these PVT systems is the electrical output. The real problem, however, is that this is a classic example of the quality of energy: the conversion efficiencies of low grade energy can be significantly higher than those for high-quality energy such as electricity. At present, few of the PVT systems use thermoelectrics to generate electricity, as the thermoelectric output is so low, but this may be a possibility in the future for ICT systems where the use of heat is not required.

4. Conclusions

To conclude, the fundamental operation and efficiency of thermoelectrics, photovoltaics and thermal photovoltaics have been reviewed. For each of the technologies, ICT applications have been described, and some of the problems have also been discussed. It is clear that these technologies are key to enable renewable sources of electricity to power ICT systems in the future if carbon emissions are to be reduced. These research fields are very active as all the technologies at present are still too expensive both in terms of capital cost and the cost per Watt of electricity generated compared to batteries and fossil-fuel-based grid electricity. Such systems are starting to be implemented where carbon taxes or subsidies can overcome the large capital investment barriers or companies are prepared to have long payback times before significant cost reductions can be achieved. As none of the technologies are close to the Carnot thermodynamic limit, there is the potential for substantial improvement in the future.

Author details

Lourdes Ferre Llin and Douglas J. Paul*

*Address all correspondence to: Douglas.Paul@glasgow.ac.uk

School of Engineering, University of Glasgow, Glasgow, UK

References

[1] Harrop., The hot applications for energy harvesting (2009). http://www.energyharvestingjournal.com/articles/the-hot-applications-for-energy-harvesting-00001247.asp

[2] H.J. Goldsmid, *Introduction to Thermoelectricity* (Springer-Verlag, Berlin Heidelberg, 2010)

[3] D.M. Rowe, *Thermoelectrics Handbook: Macro to Nano* (CRC Taylor and Francis, Boca Raton Florida, 2006)

[4] G.J. Snyder, E.S. Toberer, Nature Materials **7**(2), 105 (2008)

[5] J. Dismukes, E. Ekstrom, D. Beers, E. Steigmeier, I. Kudman, Journal of Applied Physics **35**(10), 2899 (1964)

[6] R. Venkatasubramanian, E. Siivola, T. Colpitts, B. O'Quinn, Nature **413**(6856), 597 (2001)

[7] A.I. Boukai, Y. Bunimovich, J. Tahir-Kheli, J.K. Yu, W.A. Goddard, J.R. Heath, Nature **451** (7175), 168 (2008)

[8] G. Joshi, H. Lee, Y.C. Lan, X.W. Wang, G.H. Zhu, D.Z. Wang, R.W. Gould, D.C. Cuff, M.Y. Tang, M.S. Dresselhaus, G. Chen, Z.F. Ren, Nano Letters **8**(12), 4670 (2008)

[9] Y. Ma, Q. Hao, B. Poudel, Y. Lan, B. Yu, D. Wang, G. Chen, Z. Ren, Nano Letters **8**(8), 2580 (2008)

[10] Q. Zhang, J. He, T.J. Zhu, S.N. Zhang, X.B. Zhao, T.M. Tritt, Applied Physics Letters **93** (10), 102109 (2008)

[11] T.C. Harman, P.J. Taylor, M.P. Walsh, B.E. LaForge, Science **297**(5590), 2229 (2002)

[12] X.W. Wang, H. Lee, Y.C. Lan, G.H. Zhu, G. Joshi, D.Z. Wang, J. Yang, A.J. Muto, M.Y. Tang, J. Klatsky, S. Song, M.S. Dresselhaus, G. Chen, Z.F. Ren, Applied Physics Letters **93** (19), 193121 (2008)

[13] T. Harman, M. Walsh, B. laforge, G. Turner, Journal of Electronic Materials **34**, L19 (2005)

[14] Micropelt (2016). URL http://www.micropelt.com/

[15] C. Watkins, B. Shen, R. Venkatasubramanian, in *Thermoelectrics, 2005. ICT 2005. 24th International Conference on* (2005), pp. 265–267

[16] J. Yang, F.R. Stabler, Journal of Electronic Materials **38**(7), 1245 (2009)

[17] G. Pearson, W.H. Brattain, Proceedings of the Institute of Radio Engineers **43**, 1794 (1955)

[18] S.M. Sze, *Physics of Semiconductor Devices* (John Wiley and Sons, New York, 1981)

[19] E. Bequerel, Compter Rendues **9**, 561 (1839)

[20] W.G. Adams, R.E. Day, Proceedings of the Royal Society A25, 113 **A25**, 113 (1877)

[21] W. Shockley, The Bell System Technical Journal **28**, 435 (1949)

[22] D.M. Chapin, C. Fuller, G. Pearson, Journal of Applied Physics **25**, 676 (1954)

[23] Y. Chevalier, F. Duenas, in *Proceedings of 2nd European Conference on Photovoltaics Solar Energy*, vol. 2 (1979), pp. 817–823

[24] Photovoltaics report (2016). URL https://www.ise.fraunhofer.de/de/downloads/pdf-files/aktuelles/photovoltaics-report-in-englischer-sprache.pdf

[25] R. Pierret, G. Neudeck, *The P-N Junction Diode* (Addison and Wesley, Reading MA, 1994)

[26] W. Shockley, H.J. Queisser, Journal of Applied Physics **32**(3), 510 (1961)

[27] M.A. Green, K. Emery, Y. Hishikawa, W. Warta, E.D. Dunlop, Progress in Photovoltaics: Research and Applications **23**(1), 1 (2015)

[28] T. Chow, Applied Energy **87**(2), 365 (2010)